# Global Warming

**Recent Titles in**
**Historical Guides to Controversial Issues in America**

# Global Warming

Brian C. Black and Gary J. Weisel

Historical Guides to Controversial Issues in America

 GREENWOOD

AN IMPRINT OF ABC-CLIO, LLC
Santa Barbara, California • Denver, Colorado • Oxford, England

**Library of Congress Cataloging-in-Publication Data**

Black, Brian, 1966–
    Global warming / Brian C. Black and Gary J. Weisel.
        p. cm. — (Historical guides to controversial issues in America)
    Includes bibliographical references and index.
    ISBN 978-0-313-34522-7 (hard copy : alk. paper) — ISBN 978-0-313-34523-4
(e-book) 1. Global warming.    2. Global warming—Government policy—United States.
3. Environmental policy—United States.    I. Weisel, Gary J.    II. Title.
    QC903.B53    2010
    363.738'74—dc22        2010007137

ISBN: 978-0-313-34522-7
EISBN: 978-0-313-34523-4

14  13  12  11  10    1  2  3  4  5

This book is also available on the World Wide Web as an eBook.
Visit www.abc-clio.com for details.

Greenwood
An Imprint of ABC-CLIO, LLC

ABC-CLIO, LLC
130 Cremona Drive, P.O. Box 1911
Santa Barbara, California 93116-1911

This book is printed on acid-free paper ∞

Manufactured in the United States of America

*To Nora, Sophia, Ben, Sam, and the generation that needs to find answers*

# Contents

# Introduction

Any discussion about global warming forces one to consider the possibility that human societies, in the course of what we consider the normal pursuit of economic opportunity and a rising standard of living, are causing a potentially dangerous and even catastrophic change of the Earth's climate. On first thought, this possibility seems so incredible in its scale that many people reject it. Could humans, one species in millions, carry such a burden for Earth? It seems the equivalent of a single ripple in the surface of a pond able to become a churning torrent that empties the entire body of water on to the shore.

And yet, as one considers the remarkable story of humans as Earth's exceptional species, the idea comes to seem less than surprising. Consider such a possibility, for instance, as you tour the Grande Galerie de L'Evolution in Paris, France. You might even come to see humans' ability to impact Earth as entirely likely.

The Grande Galerie begins similarly to most natural history museums: with display after display of different life forms. Two floors of photographs, models, dried plants, and stuffed animals attest to the astonishing diversity of the living world in both the terrestrial and marine environments. Although the samples are carefully organized according to accepted classification systems by the museum's curators, the sheer diversity of critters residing on Earth doubtlessly overwhelms the typical visitor. The Grande Galerie saves its knockout-punch for the third floor.

Within the general heading "Man's Role in Evolution," the Grande Galerie's third floor exhibition reminds us that humans are not separate from evolution—not simply plopped down in the midst of a living world generated around us for our benefit. One of the first exhibits places the visitor on Earth about 11,000 years ago, at the time when human beings turned from hunting and gathering to living in more concentrated agricultural settlements. During this time, *Homo sapiens* started to practice animal and plant domestication and what 19th-century British scientist Charles Darwin would later call "artificial selection." To favor certain morphologies and behaviors, they made choices in food, environments, and modification of natural surroundings. This favoring of certain characteristics by humans usually led to the elimination of others. Within a few generations, domesticated animals and plants—those chosen by human residents—thus came to represent a relatively narrow range of genetic diversity compared to that of their wild ancestors. As humans applied their impressive minds and ingenuity, their choices and cultural patterns had significantly impacted the population of the entire planet.

The next exhibit in the Grande Galerie's third floor zooms forward in time to the great voyages of discovery in the 15th century powered by humans' ingenious use of wind power. Trade networks, cultural interaction, and war became expanding portions of the human condition, and with each trek, explorers attempted to bring their own species of animals and plants from the Old World to the New, and vice-versa, a process often referred to as the Colombian Exchange. Many of the transplants could not survive in their new environments. Others, however, were so successful that they flourished and destabilized their new host ecosystems and even endangered indigenous species.

By now, the exhibit's overall point is emerging: Just as humans are participants in the lives of Earth's other species, our actions might also impact them. Thus far in the exhibit, human patterns have affected specific species directly. Of course, the scale and scope of human impacts can be much greater, even to have sometimes modified the very course of natural history in a region. For instance, the next exhibit stresses that in creating new communities and settling new lands, human beings also transformed basic patterns on the natural landscape. From the Neolithic period, in which humans cleared the land of trees to practice agriculture, to the industrial age, in which large urban areas were developed and encouraged enormous population growth, the ecological landscape of countless animal and vegetable populations has been modified and carved up, forcing genetic modifications. A further exhibit underlines the effect of pollution on animals and plants, including industrial emissions, the production of synthetic (made only by humans) substances that the natural environment cannot recycle on its own, and artificial chemicals, such as pesticides, which become concentrated in the fatty tissues of animals and threaten the continuity of species. The impact of human activity reaches the molecular level in species not even directly interacting with it!

George Perkins Marsh, the famous statesman from Vermont in the 1860s and one of the first voices of American conservation, famously stared into the development and advancement that most Americans blindly saw as "progress" and stated: "Man is everywhere a disturbing agent" (Marsh, 1965, 37). It is a lesson that visitors to the third floor of the Grande Galerie de L'Evolution relearn as the exhibition shows them that human beings have had a significant effect on the Earth's flora and fauna (including extinctions) both through direct action on living beings themselves (through domestication and transfer) and through indirect action by altering the environment (through changes in land use and pollution). Ecologists have demonstrated that desertification, shifts in forest types or river flow, and even "dust bowls" are known to have been caused in large part by human living patterns. Is it so unbelievable, then, that the scale of human impacts could be greater yet? As science and satellites provide new insight and information, could humans' impact be seen to have occurred on a planetary scale? Not only does it seem believable, but after touring an exhibit such as this, one will feel certain that a growing number of humans with an increasing level of technology have wrought planet-wide implications for Earth.

Global warming may represent a human intervention of a larger scale than ever before, but it is not different in kind. Like the surprised museum visitor upon reaching the third floor of the Grande Galerie, we are brought by global warming face-to-face with the possibility that humans might not have as much control over the natural world as we hoped, and that the stewardship of our environment has proven to be a more difficult thing than we ever imagined. It can truly be an overwhelming proposition—and that is just how it should be.

In 21st-century American culture, the term *global warming* has gained a widespread vernacular understanding that includes much more than its literal meaning. In its more expansive forms, global warming means everything from a recent rise in average global temperatures, to the identification of human activity as the cause of this trend, to grim forecasts in which human beings face huge environmental changes and, finally, to all sorts of specific recommendations about how we must change our way of life to respond to the problem. All of this baggage is an important part of our story. A reference book such as this one seeks to interest the reader and to keep him reading, but it must also seek to be a reasonably objective conduit of useful facts and ideas. As such, we must avoid overheated rhetoric, although we may on occasion need to report on such rhetoric from others.

At the outset, we should clear up a few terminological confusions. Although the terms *global warming* and *climate change* have been used interchangeably by many journalists and politicians, the terms often are distinct. The term *climate change* is sometimes used to indicate "climate variability," referring to the myriad of variations that the Earth has undergone—from the Cryogenic Age, during which our planet was covered entirely by ice, to the Mesozoic Age, when dinosaurs roamed the Earth even at its poles. The climate science community generally restricts the meaning of *global warming* to just two things. The first part hinges on the question about recent trends in average global temperature: are these best described as "down," "up," or "no change?" During the 1970s, there was considerable doubt about this most basic question; some scientists even worried about a possible cooling trend. Since the mid-1980s, however, the international scientific community has agreed that temperatures have trended upward. This is the first and most certain meaning of *global warming.*

The climate science community now attaches a second meaning to the term, connected with the hypothesis that human activity has caused the rising temperature trends of the last three decades. Scientists first considered the possibility of anthropogenic global warming during the late 19th century. For the next 50 years, it was entertained as little more than an interesting speculation. In the 40 or so years after World War II, however, knowledge about the Earth's past climate and about fundamental physical processes affecting the climate improved greatly. By 1990, the year of the first report by a large group of respected climate scientists known as the International Panel on Climate Change (IPCC), a reasonably strong consensus had developed that global warming was triggered by an increase in greenhouse gases, and that this, in turn, was due largely to human activity. Although questions continue to linger today about details of this conclusion, the IPCC says in its fourth report, of 2007, that it is made with "very high confidence." This is the second part of the term *global warming,* although it is still sometimes contested.

In this book, we present many of the historical arguments regarding both parts of the term *global warming.* As we will see, going past the simple claim of rising global temperatures to the idea that these changes were caused by human activity was difficult enough—it required extensive scientific knowledge and a worldview that placed humans as reliant on Earth's resources, not necessarily in full control of them. Taken further, "global warming" becomes even more controversial because it must be based, in large part, on conjecture.

First is the problem of how to forecast global warming for the future—a hundred years from now. Scientists first express such forecasts in terms of $CO_2$ concentrations and temperatures, but then must translate these results to the likely effects that such levels will have on human society. Here, the limits of scientific knowledge become strained. As we will see, forecasts with computer models have predicted everything from merely inconvenient shifts in human affairs to cataclysmic upsets of our society and economy. Good science can be found throughout this range of possible outcomes. The unfortunate thing is that, by the time we know which outcome is the true one, it will almost certainly be too late to act.

Therefore the issue leaves policymakers, scientists, and citizens with a profound series of questions. In the face of uncertainty regarding our forecasts for global warming, how much political will do the American people have to act on the problem? What are the risks to our natural environment if we do nothing or next to nothing? What are the risks to our way of life if we do too much?

As we will explore in this volume, the question of what to do—the mitigation of global warming or adaptation to it—is one of the most highly politicized and difficult decisions facing our generation. Some will see specific proposals as hindrances to American security and prosperity, whereas others will see them as a golden opportunity for greater efficiency and new business. But who is right? In this book, we cannot hope to answer such a question but instead hope to leave the reader more able to engage in this crucial debate as a responsible member of a democratic society.

Chapter 1 offers a layperson's introduction to essential scientific information. It begins with a description of some of the most important climate processes. A quick review of Earth's geological and climate history follows, which seeks to exemplify some of the climate processes just reviewed. A third section gives a sketch of how climate has affected the evolution of *Homo sapiens* and the history of early human civilizations. Finally, we give a first statement of the hypothesis of anthropogenic global warming.

Chapter 2 gives an historical account of how scientists pieced together information about past climate and various climate processes during the 20th century to reach the conclusion that the Earth's climate was warming and that this was due mainly to the release of greenhouse gases by human activity. We describe this agreement by most of the world's climate scientists as a "consensus." In using this word, we liken it to Thomas Kuhn's idea of a "scientific paradigm." For example, in establishing Newtonian mechanics as a scientific paradigm during the 18th century, a community of researchers shared a new constellation of beliefs, including a set of laws (Newton's Laws), calculational methods (calculus), ontological commitments (contact forces and action at a distance), and experimental procedures. The word "community" is especially important. Science for Kuhn is necessarily a social activity that involves a significant number of researchers; there is no such thing as a paradigm of one.

Although Kuhn and the historians of science who came after him have never sought to give a precise definition for a scientific paradigm, we will say simply that the international community of climate scientists, many of whom have participated in writing reports for the IPCC, have formed such a scientific paradigm, or, as we will say, consensus.

Chapter 3 rolls back the clock and asks the question: "If the hypothesis of anthropogenic warming is correct, then how did we get here?" How did human beings have such a widespread effect on the natural environment that global warming became even a plausible hypothesis? The short answer to this question is "the Industrial Revolution," but we will lead the reader through many of the remarkable changes that modern industry and economics have worked on the Earth's land, atmosphere, and oceans while offering developed nations a standard of living unthinkable in prior generations. This standard of living, ironically, provides humans in the 21st century with the intellectual ability and technological innovations to comprehend and chart the human impacts on Earth.

Chapters 4 and 5 next look at the national and international responses to the human change of the environment generally and global warming more specifically. Chapter 4 traces how rising cultural and moral awareness in the United States around a number of specific issues built the modern environmental movement. It then shows how this environmental awareness provided a backdrop, of both public concern and political contention, for the issue of global warming. The chapter closes with a portrait of America's ambivalent response to global warming: on the one hand drawing on a rich environmental ethical legacy but on the other demonstrating nervousness about endangering its traditional strength of market capitalism and limited government. Chapter 5 turns to the international context and recounts how the issue of global warming eventually made a stronger impression on other nations, especially those in the European community. It also recounts how this led to the hugely significant scientific and advisory work of the IPCC, as well as the first international treaty aimed at reducing the emission of greenhouse gases, the Kyoto Protocol.

As we reach the present day, it becomes more difficult to be objective. Chapter 6 reviews a number of specific responses to the problem of global warming, often at the state, city, and community levels. Initially, the developments happened abroad, but a remarkable grassroots effort ensued in the United States, despite reluctance by the federal government, to create policies and practices that would help to mitigate or temper rising temperatures. Many of these examples give us the hope that "the answer" to global warming and the unavoidable uncertainty that surrounds it will come from as many "win-win" initiatives as possible—initiatives that not only mitigate or adapt to the changes brought by global warming but also benefit our human society and economy in other ways, through the conservation of resources and more economical practices.

Finally, in the Epilogue, we summarize where we believe, or perhaps hope, that the attitudes of the public, business, and government are leading: toward a greater and more practical environmental awareness and a willingness to act prudently on the challenge of global warming. It would make a hopeful final portion of the exhibition at the Grande Galerie de L'Evolution in Paris and, therefore, also a promising close to our book. Perhaps this final section will be of use and interest to the reader, even though events will have no doubt raced ahead of us by the time this book reaches your hands.

It's up to you!

# 1

# What We Know: A Brief History of the Earth and Its Climate

The hypothesis of anthropogenic global warming hinges on the claim that natural causes alone cannot account for current climate trends but that human activity must be factored in (especially the burning of fossil fuels and deforestation). Therefore, reviewing an outline of the natural history of the Earth is a great aid to understanding the hypothesis of anthropogenic warming. In this chapter, we first consider the major climate processes and ways that the climate record has been established. We then turn to a review of geological and climate history, with an eye to specifying the likely causes of past changes in climate. Finally, we consider how human history might have been influenced by climate change in the past, as well as the hypothesis that human activity has caused the warming trend of the late 20th and early 21st centuries.

## CLIMATE PROCESSES

### Solar Radiation

All of the climate processes discussed in this section begin far from Earth at the sun, which behaves very nearly like a blackbody (meaning that it is an excellent absorber and emitter of radiation). Its surface burns at about 5,500 degrees Kelvin, putting out about 46 percent of its light in the visible spectrum (400 to 700 nm), about 47 percent in the infrared ranges (long wavelengths), and the remainder in the ultraviolet region (short wavelength). Of course, the sun does not emit radiation at a perfectly constant rate. For one thing, the sun has increased its overall brightness by 25 percent during the lifetime of the Earth (about 4.5 billion years). In addition, it is subject to periodic changes in sunspots and solar flares. Sunspots are regions of the sun's surface (up to 2%) where the magnetic field increases by hundreds of times above typical values. Because the interior of these regions radiate less energy than usual, they appear to astronomers as dark spots, from $10^3$ to $10^5$ kilometers in diameter (larger than the Earth!). Around the sunspots, however, are bright regions called faculae. The relative brightness of the faculae is the reason that the sun's overall radiation increases when sunspots are more numerous.

In 1843, the German astronomer Heinrich Schwabe noted that when the number of sunspots visible each day are recorded and the average taken for each year, the number of sunspots shows a cycle of about 11 years (the accepted mean is now 11.2 years). On top of that periodicity are other longer cycles, some on the order of centuries. Two previous sunspot minimums of particular importance are the Maunder minimum from about 1650 to 1700, near the middle of the Little Ice Age and the Dalton minimum from about 1790 to 1830 (Burroughs, 31; Bryant, 18).

### Earth's Orbit around the Sun

Although changes in solar activity can account for some of the changes in Earth's climate over relatively short time scales, changes in the Earth's orbit are significant for larger time scales. The simplest facts will be familiar to the reader: the Earth's orbit around the sun is slightly elliptical, with an average radius of about $1.50 \times 10^8$ kilometers. Because the Earth is tilted at an angle of about 23 degrees, its different regions go through seasons. In the Northern Hemisphere, the winter solstice (the shortest day) is December 22, and the summer solstice (the longest day) is June 22.

One of the important causes of the Earth's ice ages concerns the slight fluctuations in its orbit around the sun and, therefore, slight variations in the distribution of solar radiation across the Earth's surface. Milutin Milankovitch's search for an explanation of the ice ages (see chapter 2) showed that the Earth's orbit around the sun has three long-term cyclical variations, each of which leads to variations in the amount of solar energy reaching the Earth (the insolation). The first concerns the Earth's orbital eccentricity (how elliptical the Earth's orbit is), which has two periods of 100,000 and 400,000 years. The second cycle concerns the Earth's obliquity (the angle that the Earth's axis makes with the plane of its orbit) which is roughly 41,000 years. The third cycle concerns the precession of the Earth's axis (the orientation of the Earth's axis relative to its elliptical orbit), which has a period of 21,000 years. The Milankovitch cycles have proven of great use in explaining the rhythm of glacial-interglacial periods within the current Ice Age (starting about 2.6 million years ago) but are less useful in explaining the onset of the four or so major ice ages on the order of hundreds of millions of years ago.

Because these orbital changes do not significantly change the total amount of energy reaching the Earth, but instead its distribution over the Earth's surface, it seemed unlikely that they, in and of themselves, could account for the ice ages. Therefore, in addition to the orbital changes, Milankovitch's theory included the idea that a feedback mechanism associated with snow and ice at the Earth's poles would encourage the trend started by the initial change of the Earth's orbit. Polar ice, in the oceans and on land, has a high reflectivity, or "albedo." Therefore, if a significant amount of ice melts during a warming period, then the ice will be replaced with land. Because the land has a lower albedo than ice, it will absorb more energy of the sun's rays. This, in turn, will contribute to further warming, touching off a "positive feedback loop" (meaning that it contributes to the original change rather than offsetting it). This works the other way as well. A relatively small decrease in global temperature can lead to a somewhat greater buildup of snow and ice, thus increasing the reflectivity (or albedo) of the poles. In turn, the poles will absorb still less of the sun's overall energy and therefore lead to greater cooling (Houghton, 85; Van Andel, 90).

## Earth's Atmospheric Composition

The well-known term *greenhouse effect* refers to the tendency of the Earth's atmosphere to admit the sun's radiation but partly retain the low-frequency heat radiation. To some degree, the term is misleading. A greenhouse stays warm because the glass windows stop heat from escaping via convection, the movement of air. By contrast, the Earth's greenhouse effect does not operate because the upper layers of the atmosphere stop the movement of the lower layers like the pane of glass. Instead, the greenhouse gases in the Earth's atmosphere present a partial barrier to the transmission of low-frequency electromagnetic radiation.

After the Earth's surface absorbs solar radiation (as does its atmosphere and clouds), the Earth, like any other heated body, emits radiation. Because greenhouse gases are partially opaque to low-frequency heat radiation, some of this escaping heat energy is captured by the troposphere (the lowest layer of the atmosphere) and then radiated back to the Earth's surface, thus warming the planet (see the next section for details). It is important to realize that the greenhouse effect is an essential part of life on Earth; without it, our planet would be covered with ice. By contrast, the planet Venus has the opposite problem. Because the Venusian atmosphere has an enormously high concentration of carbon dioxide (nearly 97%), the planet is terrifically hot (about 460 degrees C).

Assuming for the moment, zero humidity, the Earth's atmosphere is currently composed largely of nitrogen, at about 78.1 percent, and oxygen at about 20.9 percent. This leaves only about 1 percent to be accounted for. Because the concentrations of the remaining gases are so low, they are expressed in parts per million (ppm). The concentration of argon is somewhat more than 9,000 ppm and therefore accounts for the lion's share of the remaining 1 percent. The last 0.1 percent of the Earth's atmosphere has a hugely important effect on the planet's climate. This is because the remaining gases include a number of greenhouse gases, including carbon dioxide, methane, nitrous oxide, ozone, and a family of human-made chemicals used in refrigeration called chlorofluorocarbons (CFCs). Carbon dioxide ($CO_2$) is the most significant with a current concentration of about 380 ppm (up from about 280 ppm around 1800). The next most significant gas is methane ($CH_4$) at about two ppm. The fact that the relative concentration of methane is much lower than that of $CO_2$ should not leave the impression that it is much less important. In fact, each molecule of $CH_4$ is more efficient than $CO_2$: currently, methane contributes about one-third the present greenhouse effect compared to that of $CO_2$, even though it is about 1/200 in concentration. Of course, at any finite humidity, a further greenhouse gas, water vapor, is introduced.

Ozone and CFCs present a complicated case, although their overall importance is less than $CO_2$ and $CH_4$. The well-known problem, that CFCs deplete ozone in the stratosphere and thereby allow more ultraviolet radiation to reach the Earth's surface, is a separate problem from the greenhouse effect; however, it is also true that both ozone and CFCs are greenhouse gases. Therefore, CFCs have a double environmental danger: by increasing the greenhouse effect, they deplete the ozone. At the same time, however, because CFCs deplete ozone, their overall greenhouse effect is partially counteracted by the depletion of ozone. That said, ozone depletion is not an acceptable result, as ultraviolet radiation is dangerous to virtually all forms of terrestrial life.

We will take this opportunity to introduce another important constituent of Earth's atmosphere: aerosols, huge collections of airborne particles. These can influence the climate either by reflecting solar radiation (resulting in cooling) or by absorbing radiation (resulting in warming). Aerosols also affect the climate by serving as the nuclei on which water vapor condenses, thus forming clouds. The overall effect of any particular aerosol depends on its composition and its height in the atmosphere. Their particle sizes vary from as small as 0.001 micrometers to as big as 10 micrometers. Also, they can stay in the Earth's atmosphere as briefly as hours to as long as years. One of the most common types of aerosols is due to volcanic eruptions. Volcanoes release huge quantities of dust and gases. One of the gases, sulfur dioxide, forms different particles through chemical reactions. The particles hang in the lower stratosphere (around 10 km in altitude) for several years and tend to stop some of the incoming solar radiation before it gets to the troposphere, thereby leading to cooling. Another common type of aerosol is due to industrial emissions, often referred to as smog. These are relatively small and short-lived particles (on the order of days) and tend to have localized cooling effects (Bryant, 22; Houghton, 50).

## The Carbon Cycle

As the most important greenhouse gas, carbon dioxide is of special concern. The concentration of $CO_2$ in the Earth's atmosphere is intimately tied to the carbon cycle, in which carbon dioxide enters the atmosphere through various "sources" and leaves through various "sinks" (or reservoirs). At any given time in Earth's history, there has been a complicated balance between sources and sinks of $CO_2$, giving rise to a fairly narrow range of atmospheric concentrations. Some of the sources of $CO_2$ are intimately tied to the sinks. The most important example of this involves the Earth's many oceans, lakes, and rivers. Because carbon dioxide can be dissolved in water (recall that soda pop gains its fizz through the addition of $CO_2$ gas), every body of water can act as a reservoir for $CO_2$. Concentrations of $CO_2$ that are relatively high in water, however, can act as a source of $CO_2$ by releasing the gas back to the atmosphere. Whether bodies of water absorb or release $CO_2$ (particularly during the breaking of waves) depends on the relative concentrations of $CO_2$ and the chemical laws that determine the equilibrium. For a change of 10 percent concentration of $CO_2$ in the atmosphere, the concentration in water solution changes by about a tenth of this, which is called the Revelle factor.

Other important natural sources of carbon dioxide include respiration by human beings and animals, the decay of plant and animal matter, fires, and volcanoes. These are balanced by other natural sinks of $CO_2$. One of the most important sinks is photosynthesis in plants. Plants produce sugars and other organic compounds by using the energy of sunlight. As part of this process, they take in $CO_2$ and give off oxygen. Of course, respiration and photosynthesis also take place in the Earth's oceans, lakes, and rivers.

There are a few relatively long-term reservoirs of carbon. The longest is probably fossil fuels. Most of the coal that we are now burning was laid down more than 300 million years ago, during the Carboniferous period (see the Phanerzoic Eon). Other long-term reservoirs involve the oceans.

Usually, $CO_2$ stays near the ocean surface and can soon be reevaporated to the atmosphere. There are two processes, however, in which $CO_2$ gets taken out of circulation for hundreds or thousands of years. In one, the solubility pump, some of the $CO_2$ slowly makes its way to the ocean depths due, in part, to circulations like the Gulf Stream. There, the higher pressures and lower temperatures enable relatively high concentrations of $CO_2$. A somewhat similar process is called the biological pump, in which carbon is transferred to the bottom of the oceans after the death of marine life. The life spans of various plankton, fish, and other animals are relatively short. When they die, about one percent of their bodies (containing carbon) end up on the ocean bottom where they are removed from circulation for as much as millions of years. The biological pump is thought to be a source of positive feedback to climate change. For example, during the cooling periods before ice ages, it is thought that the upper layer of the oceans undergo convective cooling. This mixing of surface waters would increase biological activity to somewhat greater depths and therefore increase the number of living (and dying) animals, thus hastening the biological pump.

Although $CO_2$ levels varied greatly in the Earth's distant past (even going as high as 1,500 ppm in the early Carboniferous period), they were remarkably constant before the Industrial Revolution (beginning in the late 18th century). Whereas the global concentration of $CO_2$ remained within 20 ppm of 280 ppm about 10,000 years before the Industrial Revolution, it has risen in the last two centuries to about 380 ppm. The famous "Keeling curve," summarizing measurements of $CO_2$ concentration over the last 50 years (see chapter 2), leaves virtually no doubt that concentrations have gone up dramatically. In fact, other information (from ice cores and lake cores) indicates that 380 ppm is a higher atmospheric concentration of $CO_2$ than any known for the past 650,000 years. It appears likely that the increase of the last 200 years is mostly attributable to human activities.

A number of sources of $CO_2$ resulting from human activity are not now, and perhaps never can be, balanced by natural reservoirs. These include the burning of fossil fuels such as coal, oil, and natural gas. Burning these fuels does not create new carbon but merely removes it from some particularly effective reservoirs and releases it to the atmosphere. How long the new $CO_2$ stays in the atmosphere is unclear. Based on computer models, about half of it can be removed in about three decades by other natural reservoirs, especially the oceans. But the second half appears to take considerably longer, the last 20 percent or so taking many centuries. It is estimated that all types of human activity now release about 8.0 giga tons of carbon to the atmosphere each year. It is estimated that only about 2.0 giga tons of this carbon is taken out of circulation by the oceans.

This is not helped by the fact that human beings have interfered with one of the most important reservoirs of $CO_2$ by deforestation, which lowers the number of plants and therefore the level of photosynthesis. About half of deforestation is due to slash and burn farming, in which forests and woodlands are cut down and burned (which itself releases $CO_2$) in order to create fields for agriculture and livestock. Further types of deforestation are due to the logging industry and cattle ranching. In a recent ironic development, the expanded production of vegetable oils (such as soy, corn, and palm) as an alternative to fossil fuels has led to the widespread clearing of land and, therefore, a significant release of $CO_2$. (Deforestation also causes other environmental problems including animal extinctions, desertification, and the displacement of indigenous people.) Still more $CO_2$ is released by industries, such as the cement industry, that rely on certain chemical reactions that produce $CO_2$ (Houghton, 34).

## Global Radiation Budget

At any given time, there must be a balance at the top of the Earth's atmosphere between the incoming solar radiation and the outgoing long-wavelength waves that the Earth radiates back to space. This puts the Earth at some average global temperature. When any aspect of the Earth's radiation balance shifts over a period of time—for example, as a result of changes in solar radiation or greenhouse gasses—the Earth will reach a new balance point and a new average global temperature.

The average intensity of the solar radiation incident at the top of the Earth's atmosphere is about 345 Watts per square meter ($W/m^2$). This takes into account the curvature of the Earth's surface and the fact that only half is illuminated at any moment. When we follow the sun's radiation, we find that it goes to three general places: the clouds, the atmosphere, and the Earth's surface. To make the following discussion easier to follow, we use approximate numbers. Our objective is to show the most basic relationships in the Earth's radiation budget.

About 255 $W/m^2$ of incoming solar radiation interacts with the clouds and atmosphere. Of this, 90 $W/m^2$ is reflected back out to space, another 90 $W/m^2$ is transferred in turn to the Earth's surface, and 75 $W/m^2$ is retained by the clouds and atmosphere as heat. Meanwhile, the other $345 - 255 = 90$ $W/m^2$ of incoming solar radiation interacts with the Earth's surface directly. Of this, 15 $W/m^2$ is reflected back out to space and 75 $W/m^2$ is absorbed by the Earth's surface. When we summarize these contributions from the sun's incoming radiation, we find that 105 $W/m^2$ is lost to space and, therefore, only $345 - 105 = 240$ $W/m^2$ makes it through. Of this amount, 75 $W/m^2$ is gained by the clouds and atmosphere and 165 $W/m^2$ is gained by the Earth's surface.

The energy per time that is gained from solar radiation must be balanced by the energy per time that is lost due to the reradiation of heat. Keep in mind that the Earth's surface, clouds, and atmosphere have already been heated up, and so some of the following numbers might seem to be higher than they should be. First, the Earth's surface radiates energy at a rate of 380 W/m². Of this, about 40 W/m² is lost through the "window" in the Earth's greenhouse, but a full 340 W/m² is absorbed by the clouds and atmosphere.

Let's stick with the clouds and atmosphere for now. The surface also gives energy to the clouds and atmosphere via evaporation, at a rate of 80 W/m², and convection, at 25 W/m². This means that the clouds and atmosphere gain energy at a total rate of 340 + 80 + 25 = 445 W/m². At the same time that they gain energy, however, the clouds and atmosphere also radiate heat energy: 320 W/m² goes to the Earth's surface. This absorption and then reradiation of heat energy is the greenhouse effect. The clouds and atmosphere also radiate 200 W/m² to space. If we combine the effect of heat processes on the clouds and atmosphere, we find that they lose 320 + 200 − 445 = 75 W/m², which exactly balances the gain from solar radiation!

Now let's finish the balance of accounts for the Earth's surface. We pointed out that the surface gives off 380 W/m² due to heat radiation, 80 W/m² due to evaporation, and 25 W/m² due to convection. Therefore, the surface loses a total of 380 + 80 + 25 = 485 W/m². But during this same period, it gains 320 W/m² in heat from the clouds and atmosphere. Therefore the surface experiences a net loss of 485 − 320 = 165 W/m² due to heat processes, which, once again, exactly balances the gain from solar processes!

As mentioned at the start of this section, a slight change in this energy budget will lead the Earth to reach a new balance point and a new average global temperature. Such changes are often compared by expressing each as a "radiative forcing." This is an effective average radiation (in units of intensity, W/m²) entering or leaving the top of the troposphere. The radiative forcing of each change calls on the Earth to redo its energy budget such that there is a corresponding balance in heat loss. A positive forcing (for example, due to the increase of greenhouse gasses) results in an effective increase in incoming radiation. This requires the Earth to emit more heat energy, which generally results in a higher average global temperature. On the other hand, a negative forcing (due to the aerosols released by volcanoes) requires that less heat energy be emitted and therefore results in lower global temperatures (Bryant, 27; Houghton, 26).

## Atmospheric Circulation

Virtually all of the phenomena that we call "the weather" and "the climate" take place within the troposphere, which extends from the Earth's surface to about 7 kilometers above the surface at the poles and about 20 kilometers at the equator. (Recall that the layers above this are the stratosphere, mesosphere, thermosphere, and exosphere.) The most important atmospheric circulation pattern is the Hadley cell, named after its discoverer, the English scientist George Hadley (working in the early 18th century). The Hadley circulation begins with rising motion near the equator and then a flow toward the poles near the tropopause (the transition between the troposphere and the stratosphere). The circulation then descends in the subtropics at about 30 degrees latitude and then flows to the equator near the surface. As a result of the Hadley cell, the tropics are warm and wet but the subtropics are warm and dry. There are two other similar cells, one between about 30 and 60 degrees latitudes (the Ferrel cell) and one between about 60 and 90 degrees (the Polar cell). If we count both the Northern and Southern Hemispheres, this gives a total of six cells.

These six major circulation belts are related to a number of other well-known wind currents. The first is the trade winds, which occur at the Earth's surface near the equator. These are caused by the Coriolis effect, which is due to the Earth's rotation to the east. As winds from the Hadley circulation in the Northern Hemisphere descend in the subtropical regions and blow south along the surface to the equator, they are deflected (in the reference frame of the Earth) to the right of their travel, that is, to the west. The net effect is that the trade winds blow from the northeast in the Northern Hemisphere. This is reversed in the Southern Hemisphere: as winds from the Hadley cell blow north along the surface to the equator, they are deflected to the left of their travel, again to the west; therefore, the trade winds blow from the southeast. A second set of analogous winds is created at the mid-latitudes from the effect of the Coriolis force on the Ferrel cells at the Earth's surface. These are the "westerlies," which blow from the southwest in the Northern Hemisphere and the northwest in the Southern Hemisphere.

The jet streams are fast-flowing (about 40 to 70 miles per hour), high-altitude currents, again near the tropopause. Their cause is somewhat more complex than that of the surface winds and involve atmospheric heating by solar radiation, interaction with the circulation cells, and the Coriolis force. The two main latitudes for the jet streams are in between the circulation cells. In the subtropics, this is between about 20 and 40 degrees latitude. The polar jet stream varies in latitude a bit more, between about 30 and 60 degrees. Considering both hemispheres, this gives a total of four jet streams, all of which generally travel west to east. Therefore, airplanes going from Los Angeles to New York travel about one hour faster than planes going the other way. The jet streams meander somewhat as they circle the globe. This variation is an example of the Rossby waves, which are due to small variations in the Coriolis force with latitude (Burroughs, 36; Van Andel, 52).

## Circulation of the Oceans

The conveyor-belt circulation, or thermohaline circulation, of the Earth's oceans is another major determinant of the climate. One part of this circulation system, the Gulf Stream, allows the East Coast of the United States and Western Europe to have far warmer climates than they otherwise would have. For example, Scotland would have a climate more like Alaska's, because the two areas are at roughly the same latitude.

The Gulf Stream carries warm water from the coast of Central America and the Gulf of Mexico up to the Nordic Seas off of Greenland, Scandinavia, and Iceland. Because a great deal of evaporation occurs when the water is heated in the south, it is significantly saltier than water in the north. As the surface water travels from south to north and cools down, its saltiness makes it denser than the other cool water. As a result, the water sinks 3,000 meters to the ocean floor. This sinking acts as a pump that continues to bring up water from the south. The water then crawls along the ocean bottom, as the North Atlantic Deep Water, and returns to the Southern Atlantic. There, the deep water meets with a second current of even denser water called the Antarctic Bottom Water, or Antarctic Circumpolar Current, which flows around Antarctica.

A balance between the North Atlantic Bottom Water and the Antarctic Bottom Water keeps the Gulf Stream flowing along the European coast. This balance changes when the relative densities of different parts of the circulation system change as a result of an injection of fresh water (which is less dense). Changes like this are believed to have happened in the Earth's past and to be possible in the future if sufficient ice melts in the Arctic and/or Antarctic regions.

This is a good place to remind the reader that ocean circulation is intimately connected with continental drift when viewed on a long enough time scale. The Gulf Stream itself would not be possible were it not for the Isthmus of Panama, which was formed about 3 million years ago during the Pliocene epoch. This narrow strip of land stopped the flow of water between the Atlantic and Pacific oceans, forcing it to reroute. Thereafter, the Atlantic currents were forced to continue northward, becoming the Gulf Stream (Burroughs, 73; Mann and Kump, 60).

## El Niño-Southern Oscillation

There are processes on Earth that make it more difficult for scientists to pin down their understanding of its climate. One of the best examples is the special circulation process known as the El Niño-Southern Oscillation (ENSO), which is a regional system centered near the west coast of South America, linking the circulations of the ocean and the atmosphere. The most important feature of ENSO is that the tropical eastern Pacific Ocean deviates from its normal temperature. The temperature fluctuations affect weather in the Pacific but eventually impact weather patterns around the world. Although its past behavior is reasonably well characterized by extensive climate data, ENSO has proven difficult to model with computer simulations or general circulation models. There are two general problems. First, ENSO introduces periodic variations into a relatively large region of the Earth's surface, which complicates efforts to identify global climate trends. Second, assuming that we have established climate trends, it is difficult to predict how these trends will, in turn, affect ENSO. Global warming skeptics often focus on such difficulties in understanding ENSO and other such regional circulation patterns.

The ENSO cycle has two aspects: El Niño ("the little boy") and La Niña ("the little girl"). The definition of an ENSO event is when sea surface temperature departs from normal by more than 0.5 degrees Celsius across the central tropical Pacific Ocean. When this condition is met for five months or longer, it is classified as an El Niño episode (warmer water) or a La Niña episode (cooler water). In the past, the ENSO cycle has occurred at irregular intervals of two to seven years, with each episode lasting between 5 months and 1.5 years. From 1874 to 2000, there were approximately 34 oscillations of ENSO, for an average of 3.6 years for each cycle.

El Niño and La Niña episodes are, roughly speaking, the inverse of one another. During an El Niño episode, the trade winds in the central and eastern Pacific Ocean are weaker than usual. Also, there is less upsurge of cold water from the ocean depths in the eastern Pacific, and so relatively warm water is able to spread over the surface of much of the tropical Pacific. This interferes somewhat with the Humboldt current, which normally carries relatively cold water up along the west coast of South America. The most extreme results of an El Niño event include the death of marine life, rainfall and flooding in the Eastern Pacific (from the Gulf of Mexico to Ecuador), extensive drought in the Western Pacific, and hurricanes around the islands of the Central Pacific (such as Tahiti and Hawaii).

In a La Niña episode, the trade winds in the central and eastern Pacific Ocean are stronger than usual. Because there is more upsurge of cold water from the ocean depths in the eastern Pacific, the surface temperatures of much of the tropical Pacific are lower than usual. Whereas El Niño causes higher precipitation in the Eastern Pacific and the Midwestern United States, La Niña episodes tend to reduce precipitation in these areas. Whereas El Niño causes drought in the Western Pacific, La Niña leads to heavy rains in Malaysia, Singapore, and Indonesia. In addition, Atlantic tropical cyclone activity is generally increased during La Niña. During the last three decades, ENSO has favored its positive, El Nino, regime.

In this section, we have concentrated on ENSO, as it is the most important regional circulation in discussions of global warming. A second oscillation that is worthy of mention is the North Atlantic Oscillation (NAO), which affects the strength and direction of westerly winds across the North Atlantic. Like ENSO, the NAO has two phases that are roughly the inverse of one another. In normal climate, the air pressure is relatively high in the Earth's subtropical regions and low around Iceland. In NAO's positive phase, there is an even stronger subtropical high pressure center and a lower-than-usual Icelandic low. This increased pressure difference leads to colder and drier conditions over Canada and Greenland, and warmer and wetter conditions in northern Europe and the eastern United States. In its negative phase, the NAO leads to a relatively small pressure difference between the subtropical high and the Icelandic low. This results in relatively mild temperatures in Canada and Greenland and cold, snowy weather in northern Europe and the eastern United States. The period of NAO's oscillations vary, but the average is about 10 years. During the last three decades, the NAO, like ENSO, has favored its positive regime. In addition to ENSO and NAO, there are a number other important ocean-atmosphere circulation systems such as the Atlantic Multidecadal Oscillation, and Pacific Decadal Oscillation, and the Interdecadal Pacific Oscillation, the Arctic Oscillation, and the Antarctic Oscillation (Linden, 181; Mann and Kump, 90).

## EVIDENCE USED TO ESTABLISH THE CLIMATE RECORD

Direct measurements of temperature, pressure, humidity, and so on exist only for approximately the last 150 years. The further back we go, the more uncertain the readings. Various corrections must be made to take into account the location and method of measurement. For example, in the case of thermometers, where the temperature data were calculated must be considered: sea surface water, sea surface air, land surface air, or land free air (as in balloon measurements). In the past 30 years, satellite measurements have determined average global temperature. On the one hand, the satellite measurements have been convenient, as they average over the globe with relative ease. On the other hand, the measurements require special attention and, for a time, gave confusing results (as we will see in chapter 5). Other familiar devices include barometers to measure pressure and hygrometers to measure relative humidity (with dry and wet bulb temperature readings). During the past 50 years or so, complex equipment has been used to measure the concentrations of greenhouse gases (especially $CO_2$) in the Earth's atmosphere, measured in parts per million (ppm).

To go back further than 150 years, we must rely on proxy data, or indirect measurement. One of the more familiar is the study of tree rings (dendro-climatology), which gives information about temperature and also the type of carbon in the Earth's past atmosphere. Each ring corresponds to a year of growth, appearing wider when conditions are favorable and narrower when they are not. Trees are affected by many factors: temperature, precipitation, sunlight, wind, and so on. To differentiate among these factors, at least partly, scientists select tree samples carefully. A tree at upper elevations is more affected by temperature variations than precipitation. Conversely, trees at lower elevations are more affected by precipitation changes than temperature. By using dendroclimatology, it is possible to estimate local climate temperatures up to thousands of years into the past.

Samples taken from specific tree rings also give information about the type of carbon in the atmosphere at the time that the tree ring was formed. Most of the carbon is the stable isotope $^{12}C$, but a small portion of it is $^{13}C$ and $^{14}C$. Although $^{14}C$, called "radiocarbon" because it is radioactive, is always at low levels, tree ring studies show that it has been decreasing over the last few decades. This suggests that a higher proportion of new carbon in the atmosphere is from "radiocarbon dead" sources. This lends credence to the idea that the "new" carbon is from the burning of fossil fuels, since by the time the fossil fuels form, most of the $^{14}C$ has decayed.

One of the most important proxy measures of greenhouse gases and temperature has been taken from ice cores in Antarctica and Greenland. Snow builds up layer by layer, year by year, and eventually gets compacted into ice over the centuries. Small air bubbles trapped in the ice provide a record of the past composition of the atmosphere, stretching back hundreds of thousands of years. By examining the cores, scientists can use isotopes of oxygen to estimate past temperatures. In the 1950s, the Danish glaciologist Willi Dansgaard found that the ratio of $^{18}O$ in the ice to the much more common $^{16}O$ was proportional to the temperature at the time of the formation of the ice. This is because the ice was formed by snow precipitation and for reasons involving the physics of evaporation and condensation, snow at lower ambient temperatures has a lower concentration of $^{18}O$. In addition, by analyzing the tiny air bubbles trapped in the ice, scientists can determine concentrations of greenhouse gases such as carbon dioxide, methane, and nitrous oxide at different levels of the ice core and, therefore, at different times in the past.

Scientists also use sediment layers from the bottom of ponds, lakes, and oceans to extract samples of dust, foraminifera (microorganisms), and pollen to estimate water temperatures millions of years into the past. The dust can be used to help date a sediment layer, and the foraminifera are used to determine how much of the $^{18}O$ isotope they absorbed. Again, the ratio of $^{18}O$ to $^{16}O$ taken in by the microorganism is proportional to the ambient water temperature at the time the organisms were alive. Pollen grains are well preserved in the sediment layers, lasting as long as 3 million years. Because different plants produce pollen with different shapes and textures, the pollen remains indicate the numbers and types of plants growing at the time that the sediment layer was formed. Changes in such data from layer to layer allow the climatologist to infer climate changes, although, of course, with some level of uncertainty (Burroughs, 81; Lamb, 149; Linden, 123).

## THE EARLY HISTORY OF EARTH AND ITS CLIMATE

It is helpful to have a general acquaintance with the history of the Earth and its climate since it often comes up (by way of comparison) in discussions of global warming. As we go, we give capsule explanations for why some of the climate changes happened. These explanations cannot be complete, especially as some of them are in considerable doubt, but they will at least exemplify some of the processes reviewed in the previous section (Flannery, 45).

### The First Three Eons: The Precambrian

As of 2009, The International Commission on Stratigraphy recognized four eons—Hadean, Archean, Proterozoic, and Phanerzoic—and more than 70 subdivisions referred to as eras, periods, and epochs (see Table 1.1 for a portion of this). We can make short work of the first three eons, often referred to collectively as the Precambrian, and focus on the fourth for the purposes of this book.

The Hadean eon, from 4.5 billion to 4.0 billion years ago (bya), began with the formation of the Earth, as it coalesced during the formation of the solar system. At first the Earth was entirely molten, but it slowly cooled to form the basic structure of core, mantle, and crust. The Archean eon, from 4.0 to 2.5 bya, saw a great deal of volcanic activity, leading to the formation of the first continents. Temperatures were considerably lower than those of the Hadean eon, although high enough for there to be a great deal of liquid water. The concentration of oxygen in the Earth's atmosphere was quite low and, therefore, the only forms of life that developed were anaerobic (thriving in oxygen-free environments), single-celled organisms, called prokaryotes. Although these life forms were extremely simple, they were widespread and formed large colonies that left fossil remains called stromatolites.

**Table 1.1**
**Geological Periods and Major Ice Ages (The International Commission on Stratigraphy)**

using the abbreviations:
  bya for billions of years ago
  mya for millions of years ago

Hadean eon (4.5 to 4.0 bya)

Archean eon (4.0 to 2.5 bya)

| | |
|---|---|
| Proterozoic eon (2.5 bya to 542 mya) | First ice age(s) |
|   Paleoproterozoic era (2.5 to 1.6 bya) | |
|   Mesoproterozoic era (1.6 to 1.0 bya) | |
|   Neoproterozoic era (1000 to 542 mya) | |
| Phanerozoic eon (542 mya to present) | |
|   Paleozoic era (542 to 250 mya) | |
|   Cambrian period (542 to 490 mya) | |
|   Ordovician period (490 to 445 mya) | Second ice age (460 to 430 mya) |
|   Silurian period (445 to 416 mya) | |
|   Devonian period (416 to 360 mya) | |
|   Carboniferous period (360 to 300 mya) | Third ice age (350 to 250 mya) |
|   Permian period (300 to 250 mya) | |
|   Mesozoic era (250 to 65 mya) | |
|   Triassic period (250 to 200 mya) | |
|   Jurassic period (200 to 145 mya) | |
|   Cretaceous period (145 to 65 mya) | |
| Cenozoic era (65 mya to present) | |
|   Paleogene period (65 to 23 mya) | |
|     Paleocene epoch (65 to 56 mya) | |
|     Eocene epoch (56 to 34 mya) | |
|     Oligocene epoch (34 to 23 mya) | |
|   Neogene period (23 to 2.6 mya) | |
|     Miocene epoch (23 to 5.3 mya) | |
|     Pliocene epoch (5.3 to 2.6 mya) | |
|   Quaternary period (2.6 mya to present) | Fourth ice age (2.6 mya to present) |
|     Pleistocene epoch (2.6 mya to 11,600 years ago) | |
|     Holocene epoch (11,600 years ago to present) | |

The Proterozoic eon, from 2.5 bya to 542 million years ago (mya), is broken up into three eras (Paleoproterozoic, Mesoproterozoic, and Neoproterozoic). The Paleoproterozoic era (2.5 to 1.6 bya) saw a relatively sudden increase in oxygen concentrations in the atmosphere. The chain of events causing this may be the first example of life forms changing Earth's environment. Sometime near the end of the Archean eon, Cyanobacteria (blue-green algae) evolved in the oceans. This algae used photosynthesis to gain energy from the sun and excrete oxygen as a waste product. At first, the oxygen was consumed by the oxidation of the large amount of unoxidized minerals and metals on the planet. Eventually, however, there was extra oxygen available in the atmosphere for use by new forms of life. This period is often referred to as the "oxygen catastrophe" because the sudden rise of oxygen was a death sentence to many strains of anaerobic bacteria. Aerobic bacteria (which relied on oxygen) eventually dominated during this eon. During the second era, the Mesoproterozoic (1.6 to 1.0 bya), the first single-celled organisms with a nucleus (eukaryotes) developed. Toward the end of the third, Neoproterozoic era (1000 to 542 mya), the first multicellular organisms appeared (Van Andel, 297).

The second half of the Proterozoic eon contained a number of periods of extreme glaciation for which we use the term *ice age,* starting about 2200, 1200, 900, 720, and 660 mya. Rather than subdivide these into multiple ice ages, we refer to them as a "first ice age," by which we mean all of the glaciations of the Proterozoic eon. Although data about climate are relatively scarce this far back in Earth's history, it appears that two of these ice ages, starting around 720 and 660 mya, were particularly severe. The planet was entirely covered by ice, an outcome that is sometimes referred to as the "snowball Earth." Because of the severity of these two ice ages, the period to which they belong was dubbed the Cryogenian.

Many causes have been proposed for the ice ages. It is generally accepted that one important cause is the Milankovitch cycles (discussed previously), the tiny variations of the Earth's orbit leading to changes in the solar radiation received by the Earth (the insolation). But although the Milankovitch variations probably helped trigger the ice ages, they are certainly not the sole cause and must be reinforced by other factors and feedback loops. One positive feedback loop is the warming-ice albedo feedback mentioned previously. Another involves the lowering of greenhouse gases, especially $CO_2$. Although the sun was about 25 percent dimmer during the early Neoproterozoic era than it is now, the Earth was warm enough to have liquid oceans. It is therefore believed that the Earth's atmosphere must have had a relatively high concentration of $CO_2$ and supported a strong greenhouse effect. During the onset of the first ice age, $CO_2$ levels appear to have declined sharply. Although the reason for this is not entirely understood, it is thought that the Earth's oceans acted as a vast reservoir of $CO_2$.

An especially important cause of the ice ages has to do with changes in the Earth's plate tectonics. This was mentioned only briefly in the first section, as it is not itself important to the hypothesis of global warming. In the course of the Earth's history, however, shifts in the continents led to changes in the surface topography of the Earth and the circulation system of the oceans. At the start of the Neoproterozoic era, most of the Earth's landmass was contained in a supercontinent named Rodina, centered near the equator. By about 700 mya, Rodina had moved somewhat toward the South Pole and then began to break up. The creation of new ocean currents and of high-altitude regions (in which ice was less likely to melt during the summer months) are believed to be important to the most severe glaciations (of the snowball Earth). It is generally accepted that the first ice ages of the Proterozoic eon ended largely because of widespread volcanic activity, which again enriched the atmosphere's $CO_2$ levels, thus allowing the greenhouse effect to recover (Lamb, 81, 184).

## The Phanerzoic Eon

The Phanerzoic eon, beginning 542 mya, is subdivided into three eras—Paleozoic, Mesozoic, and Cenozoic—which we will discuss at greater length. The Paleozoic era (from 542 to 250 mya) is subdivided into the Cambrian, Ordovician, Silurian, Devonian, Carboniferous, and Permian periods. The Cambrian period saw a huge development in the number, types, and complexity of life forms. This Cambrian Explosion involved the diversification of the number of multicellular organisms and the appearance of new organisms such as sponges, jellyfish, and worms. The early Ordovician climate appears to have been relatively warm. At first, trilobites dominated the animal world, but new forms of life also flourished, including brachiopods, mollusks, echinoderms, and corals. Toward the end of the Ordovician period, however, a mass-extinction event (one of five that the Earth has suffered) affected many life forms and may have been triggered by the second ice age, which was relatively shorter and mild and occurred from 460 to 430 mya.

Between the second and third ice ages lies the Devonian period (416 to 360 mya, named after a county in southwestern England). Climate varied from region to region, but, on balance, the global climate was relatively warm and humid. Terrestrial life first formed during this period; fish developed legs and arthropods (insects, arachnids, and crustaceans) became well established. Plant life also developed; seed-bearing plants spread across dry land and eventually formed gigantic forests. In the oceans, primitive sharks became more numerous, and the first lobe-finned and bony fish evolved.

The Carboniferous period (360 to 300 mya) gets its name from the many beds of coal that were laid down all over the world during this time (much of which we are burning today). One reason that so many coal deposits formed is that sea levels were much lower than in the Devonian period, which allowed for the development of many lowland swamps and forests. In addition, the period saw the appearance of bark-bearing trees, which provided a great deal

of organic material for future coal beds. The ferocious growth of the new forests and plants removed a great deal of carbon dioxide from the atmosphere,
leading to a surplus of oxygen at levels nearly double what it is today. This
abundance of oxygen allowed insects to grow to titanic sizes, with millipedes
reaching six feet in length.

The third ice age started about 10 million years into the Carboniferous
period and continued through most of the Permian period (300 to 250 mya).
The most important cause was probably a lowering of $CO_2$ concentrations
as a result of the great increase in plant activity. During the early Carboniferous period, concentrations had been high, approximately 1500 ppm. But
by the middle of the period, the $CO_2$ level had declined to about 350 ppm,
which is comparable to average concentrations today. Meanwhile, average
global temperatures in the early Carboniferous period had been warm, about
20 degrees Celsius. By the middle of the period, the average temperature
reduced to about 12 degrees Celsius which, again, is similar to conditions
today. At the end of the Permian period, there was another widespread (and
still not fully explained) mass extinction, affecting more than 90 percent of
marine species and about 70 percent of all terrestrial vertebrates.

Other causes of the third ice age include the Milankovitch cycles and
changes in the Earth's landmass distribution. During the third ice age, the
supercontinent "Pangaea" (named by the originator of the continental drift
theory, Alfred Wegener) slowly formed from many disparate landmasses. By
the end of the Permian period, Pangaea stretched from pole to pole; because it
had large landmasses at the poles, it was capable of supporting vast accumula-
tions of ice. This contributed to ice age conditions via the temperature-albedo
positive feedback (Van Andel, 131).

The Mesozoic era (250 to 65 mya) is subdivided into the Triassic, Jurassic,
and Cretaceous periods; it brought the development of entirely new forms
of life. The start of the era featured a relatively warm and dry global climate
compared with the Paleozoic era. Most evidence suggests that there were little
or no polar ice caps. In these conditions, the dinosaurs flourished, some living even at the poles. The first dinosaurs developed in the late Triassic period
and were probably relatively small, bipedal predators (Eoraptor). The Jurassic
period brought the development of the Sauropods (including Apatosaurus),
and the Cretaceous period brought the Theropods (including Tyrannosaurus
Rex) and the Ceratopsidae (including Triceratops). During the Jurassic era, the
supercontinent Pangaea broke up into two large masses, becoming two super-
continents, Gondwana and Laurasia, divided near the equator (Lamb, 194).

The end of the Cretaceous period saw yet another widespread extinction, sometimes called the K-T extinction. Again, the cause of this event
is uncertain, as is the quickness of the change. Scientists have sought to

explain the K-T extinctions with one or more catastrophic events, including extensive asteroid impacts and volcanic activity. Whatever the case, by the start of the next era, most of the dinosaur species were extinct, except for certain species of birds, which today are the dinosaur's only survivors (Van Andel, 382).

The Cenozoic era (65 mya to the present) is divided into three subdivisions, the Paleogene, Neogene, and Quaternary periods. The entire Cenozoic era has been a period of relative cooling (although it is not classified as an ice age until 2.6 million years ago). One explanation for the cooling of the early Cenozoic era is that at the end of the Cretaceous period, aerosols associated with the K-T extinction event served to block incoming solar radiation. In addition, the detachment of Australia fully from Antarctica (which, in turn, had broken off from the supercontinent Gondwana) allowed the Antarctic Circumpolar Current to start, which flowed west to east around Antarctica and brought cold Antarctic water from the ocean depths to the surface.

The Paleogene period (65 to 23 mya) is subdivided into the Paleocene, Eocene, and Oligocene epochs. It is most notable for being the time that mammals evolved from the sauropsids (which date back to the end of the Carboniferous period), into a wide range of life forms. The main line of the sauropsids evolved into modern-day reptiles and birds, and the synapsid branch evolved into mammals. Although the first true mammals appeared in the Jurassic period, they only flourished in the wake of the K-T mass extinction.

The Neogene period (23 to 2.6 mya) is divided into the Miocene and Pliocene epochs. Warm conditions made a comeback during the Miocene epoch, as a result of uncovered gas hydrates (icelike solids formed from mixtures of water and gas), which released methane and carbon dioxide and therefore strengthened the greenhouse effect. During the Pliocene epoch (5.3 to 2.6 mya), however, South America became attached to North America, creating the Isthmus of Panama. The continents more or less reached their current positions. Two important ocean currents strengthened: the Gulf Stream, the warm Atlantic current flowing north on the eastern coastline of the United States, and the Humboldt current, the cold Pacific current flowing north along the west coast of South America.

The fourth ice age started about 2.6 mya, corresponding roughly with the start of the Quaternary period and the appearance of the first human ancestors (the genus *Homo*). Currently, there is some disagreement about whether the Quaternary period should be considered as part of the Neogene period or should be its own period, based on geological, climatological, and biotic considerations. We will go with the second option here and consider the Quaternary period as the third subdivision of the Cenozoic era. In this case, it is divided into two epochs: the Pleistocene (2.6 mya to 11,600 years ago), followed by the Holocene.

It is perhaps surprising that not all of the ice ages were a continuous period of cold and glaciation. Based on ice core data, the current ice age had numerous oscillations between periods of relative cold (called "glacials") and warmth ("interglacials"). The length of time of a glacial-interglacial cycle is about 100,000 years. At present, we are in the midst of a warming period in the seventh glacial-interglacial cycle during the last 650,000 years. The previous glacial period, from 90,000 to 10,000 years ago, led to glaciation across most of North America, reaching thicknesses of about 2 kilometers. In North America, these ice sheets are called the Laurentide and the Cordilleran and, in Europe, the Fennoscandian. This glacial period abated between 20,000 and 10,000 years ago, leading to the second epoch of the Quaternary period called the Holocene (from 11,600 years ago to the present), which is frequently referred to as "the long summer."

The Holocene epoch has a number of points of interest that alert us to the fact that just as an ice age can have significant variations within it, so can an interglacial. The period leading up to the Holocene epoch was, as one might expect, a gradual warming trend. This was interrupted, however, by a sudden shift to arid and cool conditions, and even a partial return of the glaciers between 12,800 and 11,600 years ago. This period has been given the name Younger Dryas because of the reappearance of the small plant *Dryas octopetala,* which thrives in cold conditions and had reappeared in the Arctic. The Younger Dryas period is believed to have been started by a reduction or shutdown of the North Atlantic thermohaline circulation, as the result of the deglaciation of North America and the sudden addition of fresh water in the northern seas. The end of the Younger Dryas appears to have taken place in a relatively short period, between 40 and 50 years. This alerts us to the surprising speed at which the Earth's climate can change. It is important, however, to add the proviso that, although the Younger Dryas has been well confirmed in Western Europe and Greenland, it is not clear if it was a global phenomenon. Once the Holocene stabilized, it demonstrated variations that had a significant effect on human civilization. These are discussed in the following section and include the Medieval Warm period, a time of relative warmth from about 950 to 1250 c.e., and the Little Ice Age, from approximately 1350 to 1850 c.e. (Burroughs, 239; Bryant, 77).

## CLIMATE CHANGE AND HUMAN HISTORY

Earth's climate influences the living patterns of every species. Human development and history, however, has a uniquely dynamic relationship with climatic variations. This assumption was not always appreciated. A number of prejudices and limits to scientific knowledge delayed serious consideration or acceptance of this interrelationship.

For one thing, until the mid-20th century, it was thought by most archeologists, historians, and scientists that the Earth's climate was relatively constant and stable (Linden, 4). Furthermore, historians and archaeologists tended to concentrate on political, social, and economic relations to explain the development of human communities. To such scholars, citing climate change (or climate variability) as a primary cause of change in human affairs ran the risk of offering a simplistic explanation for complex human relations. Furthermore, explaining the earliest cultural and biological evolution of human beings with climate change was necessarily speculative; as scholars sought to go back further into the past, it became hard enough to establish an adequate record of human history and climate change separately, much less to make causal connections between the two. During the 1960s, pioneers like climate historian H. H. Lamb explored such ideas. But work like Lamb's had to wait until recent decades, when data on the Earth's climate and its theoretical interpretation improved dramatically. Recent work has also established the possibility that the Earth's climate can change significantly in relatively short periods, even within the span of a single human generation. As these breakthroughs were made, our knowledge of human origins continued to improve and certain scientists eventually became interested in seeing what human history might look like through the lens of climate change.

An interesting theory on the interaction of climate change and early human evolution has been proposed by the paleoanthropologist Richard Potts of the Human Origins Program at the National Museum of Natural History in Washington, D.C. Potts has hypothesized that early humans did better than their closest relatives because of their ability to adapt to the sudden climate changes of the Pleistocene epoch. The genus *Homo* appeared around 2.4 mya, probably diverging from the genus *Australopithecus*. Potts finds that, during periods of climate variability, the relatively more adaptable species of *Homo rudolfensis* and *Homo ergaster* seem to have done much better than the more specialized species of *Australopithecus boesei*. Later on, certain species of *Homo* seem to have been more adaptable than others. For example, *Homo sapiens* (modern humans), which appeared in East Africa about 200,000 years ago, did much better than *Homo neanderthalensis*. The Neanderthals developed only the simplest stone tools from materials close at hand, whereas *Homo sapiens* made more varied stone tools from traded materials originating hundreds of miles away. Even the jump in brain size from *Homo erectus* to *Homo sapiens* might be connected to the relative ability of *Homo sapiens* to adapt to changing conditions (Linden, 34).

The tool-making materials left behind by prehistoric human communities have formed the basis of chronologies adopted by archaeologists. The familiar three main periods are the Stone Age, the Bronze Age, and the Iron Age, which were originally proposed in the mid 19th century by C. J. Thomsen (for organizing the collection of the National Museum of Copenhagen). The Stone Age is subdivided into the Paleolithic and Neolithic periods and extends from around 2.0 mya to 6,000 years ago. It thus accounts for most of human prehistory.

Changes in climate brought concomitant changes in ways of life for *Homo sapiens*. The last major climate fluctuation was the relatively cool and arid Younger Dryas period (from 12,800 and 11,600 years ago). This appears to have encouraged initial attempts at farming, animal husbandry, and small settlements in regions featuring permanent water sources (lakes, rivers, and springs).

After surviving at least two major glacial–interglacial cycles and the Younger Dryas period, humans benefited from the relatively temperate climate at the start of the "long summer" of the Holocene epoch. The Neolithic Age (or New Stone Age), began in the Levant (the area around the West Bank) and represented an enormous change in human culture. Much of this change depended on the warming trends, which began around 11,600 years ago and leveled off between 10,000 and 8,200 years ago. The higher temperatures and wetter conditions brought a great expansion of flora and fauna. The Mediterranean area became the center of changes leading from a culture based on hunting and gathering to one based on farming, the domestication of plants and animals, and the settlement of villages and towns. As conditions improved, farming settlements spread rapidly. Early spiritual beliefs regarding the fertility of the land and humans, as well as the connection of tribal ancestors with the land, developed in these early agricultural societies.

Even though the Holocene epoch was relatively stable compared with what came before, it nevertheless contained climate fluctuations strong enough to affect the early human settlements. The first and most important of these was a cooling around 8,200 years ago (6200 B.C.E.) that some historians describe as a "mini ice age." This appears to have been caused by the melting of the Laurentide ice sheet that released enough fresh water to slow down the thermohaline circulation. The temporary return to relatively cool and arid conditions (somewhat like the Younger Dryas but probably not as severe) made many agricultural settlements unviable, especially the ones that had developed relatively far from permanent water sources. Some communities were forced to relocate to larger and more reliable bodies of water, and others wandered with their domesticated animals in a pastoral lifestyle. Still others reverted to hunting and gathering. After about 400 years (7,800 years ago), relatively mild conditions returned (Fagan, 99).

Another cooling took place starting around 5,800 years ago and reached a peak around 5,200 years ago (3200 B.C.E.), although it was probably less substantial than the cooling of 8,200 years ago. This again had the effect of consolidating human settlements, as the somewhat cooler and drier conditions were not as favorable for widespread agricultural production. Archaeological evidence suggests that, with the resettlement and expansion of cities, human cultures eventually developed more centralized (and often warring) societies. This marked the foundation of the first civilizations and empires. Mesopotamia, the Fertile Crescent between the Tigris and Euphrates Rivers, saw the development of the Sumer, Akkadian, Babylonian, and Assyrian empires. In Egypt from 3500 to 3000 B.C.E., relatively arid conditions forced tribes to concentrate along the Nile River and use the yearly flood for irrigation, eventually leading to the start of Old Kingdom around 2650 B.C.E. This marks the beginning of recorded history with the invention of writing (in part to keep track of economic relations and trade).

The foundation of the earliest civilizations also brought technological innovation. The tool-making impulse of the Neolithic revolution continued and intensified as metal tools became widespread. The Bronze Age is dated between about 3300 and 1200 B.C.E., with a Copper Age sometimes added, from about 4,000 to 3,000 B.C.E., as an intermediary. The emergence of metallurgy occurred first in the Fertile Crescent. Incan civilization (areas of South America and West Mexico) arrived at copper- and bronze-based technology independently. Bronze was useful for a wide variety of applications including knives, axes, and armor. It also represented a sophisticated step in metallurgy, in which copper was hardened by combining it with tin. The Iron Age began about 1200 B.C.E. in the ancient Near East, Iran, India, and Greece but somewhat later in other regions (during the eighth century B.C.E. in Central Europe and the sixth century B.C.E. in Northern Europe). Iron metallurgy required the development of high-temperature smelting, in which iron was extracted and purified from iron ore. An even stronger alloy, steel, was developed by leaving trace amounts of carbon in the iron.

One of the earliest and clearest examples of how climate change helped bring a sudden shift in human civilization is the fall of the Akkadian Empire. This empire was established between the Tigris and Euphrates Rivers (in modern-day Iraq and Syria) around 2300 B.C.E. but collapsed after only about 100 years. The causes of this decline had been somewhat elusive. In 1978, however, Harvey Weiss, an archaeologist at Yale, excavated a lost city, Tell Leilan, in Syria. Weiss found layers of dirt from around 2200 to 1900 B.C.E. that showed no signs of human habitation and suggested that there had been a prolonged drought and agricultural collapse. He concluded it was likely that the Akkadian Empire had been ended by climate change.

Paleoclimatologists have confirmed Weiss's conclusion. One confirmation came from Peter deMenocal, of the Lamont-Doherty Earth Observatory (at Columbia University), who did work on ocean cores. He confirmed the general finding that the Holocene epoch was much less stable than previously thought and showed recurring cool periods about every 1,500 years. DeMenocal also studied core samples from the Gulf of Oman and saw telltale traces of dolomite (a mineral that is not common in the Gulf region and could only have been transported there by wind) at depths consistent with Weiss's account (Kolbert, 91).

Since Weiss's work, paleoclimatologists have identified a number of other examples of civilizations that may have been ended by climate change. For example, the decline of the Old Kingdom of Ancient Egypt may have been caused, in part, by severe drought. A drop in precipitation between 2200 and 2150 B.C.E. appears to have prevented the normal flooding of the Nile and led to widespread famine. Climate change also appears to have been a factor in the collapse of Mayan culture from about 850 to 950 C.E. (in modern-day Central America). Although many archeologists still prefer nonecological explanations (including overpopulation, slave revolt, foreign invasion, and the collapse of trade), improved knowledge of past climate has given rise to plausible explanations for the Mayan collapse, including severe drought and a failure of agricultural production (soil exhaustion and water loss).

Two periods of climate change are particularly important to the study of Western civilizations and the Northern Hemisphere: the Medieval Warm Period (about 950 to 1250 C.E.) and the Little Ice Age (about 1350 to 1850 C.E.). It is uncertain if these changes were truly global or if they affected only the Northern Hemisphere. As of this writing, consensus goes with the second option. The Little Ice Age appears to have been a period of relative cool in mainly the Northern Hemisphere, with average temperatures dropping by about 1 degree. Because of its limited geographical extent and relatively small temperature change, the Little Ice Age is a misleading name. Scientists have suggested that both of these climate shifts were caused in part by changes in the strength of North Atlantic thermohaline circulation and, in part, by changes in solar activity (recall that the Little Ice Age overlaps with the Maunder minimum in sunspot activity).

A number of examples can be given for how these climate changes affected civilizations in the Northern Hemisphere. Some scholars believe that the rise of the Norse was encouraged by the Medieval Warm Period. The Vikings took advantage of relatively ice-free seas to colonize Greenland and other northern lands and greatly extend their power. With the onset of the Little Ice Age, the Viking settlements slowly died out. The Norse colonies in Greenland vanished by the start of the 15th century as the result of food shortages and the difficulty of maintaining livestock were through the harsh winters. Of course, climate change probably cannot account for all of the Norse's problems. More traditional explanations, involving political, social, and economic problems, still have validity (for example, the Norse had gone beyond their agricultural capacity and had trouble maintaining international trade). Climate change merely adds an additional cause, the relative importance of which may never be entirely settled (Linden, 9).

There is some evidence that the Medieval Warm Period and the Little Ice Age affected the Southern Hemisphere as well as the Northern Hemisphere. For example, the Medieval Warming Period is often invoked to help explain the fall of Tiwanaku civilization, a precursor to Inca civilization. Tiwanaku civilization originated in a region near Lake Titicaca (in modern-day Bolivia) around 400 C.E. The people living in the Titicaca Basin practiced a unique farming technique known as raised-field agriculture, in which crops are cultivated on a series of raised beds, separated from one another by irrigation channels. The power of Tiwanaku civilization expanded until about 950 C.E., when evidence suggests that a great drought occurred. As rainfall decreased, cities produced fewer crops. The capital city on Lake Titicaca had somewhat more resilient production because of the raised fields, but it appears that the lack of precipitation eventually hurt production there as well. This has led many paleoecologists (including Alex Chepstow-Lusty at the Université Montpellier) to conclude that Tiwanaku civilization disappeared around 1000 C.E., in part, because of a failure of food production (Fagan, 238).

## GLOBAL WARMING: ARE WE LIVING IN THE ANTHROPOCENE?

The theory of global warming reverses the logic of the previous section. Instead of climate affecting human history, we are faced with the possibility that human activity is modifying the Earth's climate. This has led some scientists to use the term *Anthropocene* to refer to the period after the Industrial Revolution, which we might date from the use of steam engines in the late 18th century (see chapter 3). As discussed in chapter 2, the hypothesis of anthropogenic global warming developed slowly in the course of the 20th century as direct measurements showed an overall increase of global temperature and a steady increase in atmospheric $CO_2$ concentrations.

Global warming is not the first time humans have considered scenarios in which they attempt to change or inadvertently change the Earth's weather and climate. An example of the former is a series of projects that attempted to lessen the strength of hurricanes. The first of these was Project Cirrus during the late 1940s, a collaboration of various U.S. military agencies and the General Electric Corporation. One of the early attempts was in October 1947, when an airplane seeded the clouds of a developing hurricane off the east coast of the United States with about 100 kg of crushed dry ice. The hurricane changed direction and made landfall near Savannah, Georgia. It was never established that the seeding had any significant effect. Nevertheless, certain scientists and members of the public claimed that the hurricane's redirection was due to the seeding. This led to political and legal troubles and, eventually, the cancellation of the project (Cotton, 3). More than 10 years later, a second effort was attempted by the U.S. government from 1962 to 1983. Called Project Stormfury, it sought to weaken tropical storms by seeding them near their eyewall with silver iodide. An early seeding attempt was Hurricane Beulah in August 1963. The storm had an atypical structure and the airplanes made many attempts to seed it in the correct locations. Another effort in 1969 with Hurricane Debbie gave somewhat better results, showing noticeable reductions of wind speeds. After this apparently promising result, a number of further attempts were made during the 1970s, but few clear effects were observed (Cotton, 63).

An example of humans considering a scenario in which they inadvertently (and disastrously) change the Earth's climate is the "nuclear winter" hypothesis. This is the idea that a global nuclear war would lead to the release of large amounts of smoke, dust, and aerosols into the atmosphere (as a result of the explosions themselves but also the fires following them). These materials would block enough incoming solar radiation such that the Earth's land areas would cool and winter-like conditions would persist throughout the year. The hypothesis was first brought to the public eye in 1983 by the astrophysicist Carl Sagan and colleagues, in a study based on computer models. This work has been strongly criticized by a number of scientists, some claiming that it is biased science motivated by antiwar sympathies (Cotton, 203). Nevertheless, some work in this field continues. Recent studies account for current nuclear arsenals (which are somewhat smaller than they were in the 1980s) and use the most advanced climate models, which were developed by global warming researchers. One study from 2007, using the Model E global circulation model from the NASA Goddard Institute for Space Studies, found that average global surface cooling would be as much as −8 degrees Celsius and that, even after a decade, the cooling would still be about −4 degrees Celsius.

The hypothesis of global warming has differences and similarities with both of these examples. To mention a few, unlike Projects Cirrus and Stormfury, global warming concerns the climate of the entire globe and not merely the weather. (The term *weather* refers to the temperature, air pressure, and other atmospheric conditions relating to a particular time and place; *climate* is the established pattern of weather for a region based on a broad statistical sample covering many years.) At the same time, the cloud-seeding projects highlight how difficult it is for humans to bend either weather or climate to their will. In contrast to the problem of nuclear winter, global warming does not have a single cause (the use of nuclear weapons), which, if it were stopped, would fix the problem. Nuclear winter, however, does highlight the fact that avoiding a global threat requires the confrontation of difficult political problems.

As stressed in the next chapter, the discovery of global warming has required a great deal of climate data, the understanding of a large number of climate processes, and the coordination of the international scientific community. The first thing to note is that, even if we remain agnostic about cause, the weight of the evidence strongly suggests that global warming is taking place. First, the temperature record has been pored over for many years, including the instrumental record and proxy measurements. Second, extensive ice melts have been noted in the Arctic and even to some extent in the Antarctic. Sea levels have risen by about 20 cm because of two effects: the melting of the ice caps and, even more, the thermal expansion of water as a side effect of warming. Rising sea level has been particularly noticeable in communities near the water line such as low-lying Pacific islands and the Chesapeake Bay in the United States. The Netherlands has been forced to update its system of barriers and seawalls and to abandon its efforts to protect certain rural areas. Changes in animal habits have also been noticed, including mating and migration. One simple example is that certain species of butterflies have moved northward. Sometimes the animals cannot adapt and go extinct; this appears to be what happened to the golden toad, which once lived in the mountains of Costa Rica (Kolbert, 67, 120). Somewhat less certain data are offered by the incidence of extreme weather events, ranging from droughts to hurricanes. Although some doubt remains about the connection between global warming and extreme weather, it has been enough to worry insurance companies, which have suffered a higher rate of payout.

Explaining global warming presents us with the need for a revised theory of how the known processes can combine to bring about climate change. Changes in the Milankovitch cycles and solar activity, which have been important to climate change in the distant past, cannot be primary factors in the recent warming trends if, indeed, they are caused by human activity. As of this writing, the consensus within the climate community (which, nevertheless, is not shared by absolutely everyone, as we will see) is that the most important causal agent is the increase of greenhouse gases in the Earth's atmosphere resulting from human activity (primarily the burning of fossil fuels and deforestation). The strength of this effect can be expressed as a "positive radiative forcing," or the change in the equivalent additional average radiation entering the top of the Earth's atmosphere. Current estimates place the radiative forcing resulting from the increase of greenhouse gases at 2.6 W/m$^2$. As important as greenhouse gasses are, they are not alone sufficient to explain current global warming trends. One of the most important things learned in the last 50 years about the Earth's climate is that it is a delicate system full of many interlocking processes and feedbacks (Maslin, 102).

One particularly important factor, which has been concomitant with the release of greenhouse gases, is the change in land use resulting from human activity. Over the last 200 years, human beings have cleared a great deal of land and jungle for farming and habitation. This has led to a general decrease in the plant cover of the Earth which, in turn, has led to a general decrease in photosynthesis and therefore a decrease in the absorption of $CO_2$. (This effect is often counted as part of the $CO_2$ radiative forcing.)

One of the positive feedbacks (i.e., a process that magnifies the warming trend started by the increase of greenhouse gases) is changes in the reflectivity (or albedo) of the Earth's poles. As the planet warms up and melts ice at the poles, the reflectivity of the land decreases. Another positive feedback is that ice contains a great deal of trapped methane, the remnants of living organisms. As the ice melts, it releases the methane to the atmosphere, which, as another greenhouse gas, serves to increase further warming. In general, it is believed that such processes outweigh the negative feedbacks that offset the warming trend. As an example of such a process, aerosols released to the atmosphere from both natural and human activity often partially block the sun's radiation and thereby contribute to cooling (although under certain conditions, it is possible that aerosols can increase warming by absorbing radiation and then transferring it to the Earth's surface).

Putting together all of the factors mentioned here (along with others) requires the use of computer modeling. As we will see, the development and interpretation of computer models, or general circulation models (GCMs), has been the subject of much scientific and public debate. Part of the problem has to do with the nature of such models. Using the word "predictions" to describe the results of GCMs is probably too strong (although we sometimes use it for convenience). The words "projections" or "scenarios" are more appropriate because GCMs include many uncertainties regarding the available empirical data, the theory of climate processes, and the future of human social and economic conditions. Despite these uncertainties, the various models reliably demonstrate general trends. For example, current projections of the rise in average mean global temperature by the year 2100 vary from 1 to 6 degrees Celsius (Flannery, 153).

As we will see in the following chapters, the hypothesis of anthropogenic global warming is a unique challenge that pushes the scientific enterprise to its limits in both technical and organizational terms. In addition to the scientific challenges, and partly because of them, global warming also presents political, social, and ethical challenges for humanity.

# 2

# How We Know: A Brief History of Global Warming Research

This chapter outlines the development of scientific research regarding climate change and the formation of a scientific consensus during the 1980s regarding anthropogenic global warming. Our use of the term scientific consensus is similar to Thomas Kuhn's influential idea of a scientific paradigm. For Kuhn, a paradigm was a constellation of commitments shared by a scientific community regarding its object of study including acceptable types of data, agreed upon laws and models, and experimental and computational methodologies. Reaching such social agreement within a scientific community is a sine qua non of scientific study (Kuhn, 174). Against this, many onlookers tend to favor the heroic conception of scientific progress (overlapping easily enough with American individualism) in which the most important contributions are made by great geniuses. For them, research on subjects such as global warming may always seem tainted, as though issuing from a committee or from "group-think," rather than from the authentic voice of the lone truth-teller. There is little one can do in response to this except to point out that it represents an unrealistic view of the history of science and one certainly out of step with the realities of the contemporary scientific enterprise.

Our response to such questioning composes the following pages of this chapter: we will trace the evolution of the scientific thought that has functioned—brick by brick—to compose the understanding that today we refer to as *climate change* and *global warming*. In this fashion, we can properly appreciate the remarkable cast of thinkers and ideas that have over generations asked questions and found results that have led scientists to construct a robust paradigm. Although the connection between human activity and climate change remains a powerful hypothesis—actively being tested in order to prove or disprove it—it has been built on work reaching back centuries.

## THE NATURE OF THE SCIENTIFIC PROBLEM

In seeking to prove the hypothesis of anthropogenic global warming, scientists have gathered a great deal of data from many different sources. They also have found that a great many physical processes affect the Earth's climate (see chapter 1). As more pieces were added to the puzzle, scientists reached the idea that there could not be one major cause of climate change but rather a tangle of causes and feedbacks. They also came to realize that climate change might not always be as gradual as they had once assumed, but could be relatively sudden.

It was hard enough to gather climate data and to understand the various climate processes; putting all of these elements together into a model that yielded, in some sense, a predictive theory was even more difficult. The idea that the Earth's climate is the result of many linked causes, some working together and some working against one another, has both encouraged and necessitated the use of advanced computation. These needs—a massive database, a consideration of complex causation, and the resort to computer modeling—are characteristic of many modern scientific problems. Scientists studying global warming also face a different, though no less vexing, difficulty of communicating their results to others and of convincing them that the knowledge that they have gained has value in spite of its unavoidable uncertainties. A good example of this is the Intergovernmental Panel on Climate Change (IPCC), which released its fourth report in 2007. As the product of six years of work by an international team of about 800 contributing authors from about 130 different countries, the fourth report of the IPCC swelled to encompass four volumes totaling nearly 3,000 pages. Such a gigantic and complex effort is difficult to summarize in the popular media, in Congress, and even in the general scientific community.

Because of its inherent complexity, climate science contrasts with much other scientific research, especially that of earlier centuries. For example, the history of physics has involved a significant number of experiments that have been successful because they radically simplified the subject being studied. Such experiments have sometimes been called "crucial experiments," in the sense that well-focused tests clearly decide between different hypotheses. In popular parlance, such tests might be called "smoking guns." A paradigm example of this is Isaac Newton's work with prisms. To prove that prisms separated white light into its component parts, and did not merely produce amusing but otherwise uninteresting effects (as was widely believed at the time), Newton broke white light into a spectrum with a first prism and then sent a single color through a second prism, showing that the color was not further divided.

There are many other examples of "crucial experiments," although many of these begin to cast doubt on how widely the concept can be applied. One such example can be found in Albert Einstein's development of the theory of special relativity in 1905, in which the speed of light is constant in all inertial reference frames. It was once taken for granted that the Michelson-Morley experiment of 1887 had acted as a crucial experiment for Einstein. Michelson and Morley had used a high-precision interferometer in an attempt to detect the luminiferous aether, which was presumed in the late 19th century to be the medium of electromagnetic waves. Because the experiment's null result strongly suggested that there was no aether and also strongly supported Einstein's theory, many physicists and historians assumed that Einstein was aware of the Michelson-Morley result. In fact, it appears from Einstein's own reflections that he developed the special theory of relativity without any knowledge of the interferometer experiments (Holten, 279).

The idea of clear-cut proof is even more problematic in the study of complex systems such as global climate. Although an idea of the crucial experiment might be useful in deciding research regarding one piece of the global warming puzzle, the idea is not likely to be helpful in putting all the pieces together. In some ways the theory of global warming has similarities with Charles Darwin's theory of evolution. Darwin understood clearly that he did not have any single piece of evidence that proved conclusively that life forms had evolved through natural selection. Instead, *On the Origin of the Species* presented, in Darwin's words, "one long argument" to support his theory, composed of evidence taken from many sources. His evidence included variation under domestication, variation in nature, hybridization, the fossil record, the geographical distribution of species, ways in which his theory cast light on established classification schemes, and more. None of these pieces of evidence could be taken to "prove" the theory of evolution, but Darwin hoped that the preponderance of the evidence would establish the theory as being viable and worthy of further research (Ruse, 43).

Similar to the theory of evolution, the theory of anthropogenic warming relies on many pieces of information that, individually, do not and cannot prove the theory. The corollary to this is that both theories are open to piece-meal attack, as the history of both shows. Although the theory of evolution has many more years of research supporting it than does global warming, even evolution continues to be subjected to attacks from skeptics who tend to pick on individual problems or gaps in the theory. They obdurately ignore the preponderance of the evidence and instead insist on specific issues (such as gaps in the fossil record, problems with carbon dating, uncertainties about macroevolution, and so on). Proceeding in this way, it is easy to capitalize on this or that difficulty and to leave the impression that uncertainty regarding these parts of the theory casts fundamental doubt on the entire edifice.

Even many scientists play this game—especially when they are from areas of research other than the one that they are criticizing. More often than not scientists who are skeptical of evolution come from fields that are not closely related to biology, such as physics, chemistry, geoscience, and engineering. Even when researchers from the life sciences criticize or reject the theory of evolution, such critics usually do not work in field biology (where evolution is of central importance) but from other loosely allied areas. One well-known example is the biochemist Michael Behe, who has written extensively in support of the intelligent design movement. Although many of these skeptics base their criticism on important issues that are often related to areas of ongoing research, many others do little more than stubbornly repeat questions that have already been addressed many times by the scientific community.

Finally, it is important to note that global warming research relies—even more than on scientific work of the past—on the current efforts of many individuals. Newton famously said: "if I have seen further it is by standing on the shoulders of giants." And, indeed, Newton's *Principia* would have been impossible without the previous work of Galileo, Huygens, and Descartes, just to name a few. Furthermore, Newton's mechanics was confirmed and extended by the many scientists who followed in the 18th century (such as Laplace, Lagrange, and Johann Bernoulli). Much the same can be said about Darwin's *Origin* in relation to research both leading up to and confirming and extending the research program of evolutionary biology. It has always been misleading, therefore, to tell the history of science as the history of great geniuses. This danger only increases as we come to the present day, when the most challenging scientific problems require collaborative and international research to make any headway.

## EARLIEST RESEARCH (BEFORE WORLD WAR II)

The study of the Earth's history was once limited by chronologies based on a literal interpretation of the Bible. The most famous of these is the work of Anglican Archbishop James Ussher in the 17th century, which concluded that the Earth was about 6,000 years old. The study of geology and the history of the Earth became a science only during the late 18th century, during which time two contrasting theories arose, each purporting to explain how the rock layers of the Earth's surface had formed. On the one hand, the German geologist Abraham Werner and his followers (the Neptunist school) suggested that the Earth's layers had precipitated from an ocean that once had covered the Earth (a reference to the biblical deluge). On the other hand, the Scottish naturalist James Hutton and his followers (the Plutonist school) argued that the Earth formed through the gradual solidification of molten rock originating in volcanic processes. Because the rate of volcanic activity was slow and steady, Hutton was one of the first scientists to suggest that the Earth was considerably older than 6,000 years. Although scientists later realized that only one form of rock (igneous) is formed by volcanoes, Hutton's idea of uniformitarianism—that geological processes go slowly and steadily—proved influential.

One of the earliest questions asked about the Earth's climate was how to explain average global temperature. Early in the 19th century, the French mathematician and physicist Joseph Fourier noted that, while it was obviously true that the sun provided abundant energy to the Earth, it must also be true that the Earth, in turn, lost a portion of this energy back out into space in the form of infrared radiation. To explain how the Earth retained part of the energy delivered by the sun, Fourier hypothesized that the atmosphere must hold in the heat, somewhat in the way that a pane of glass retained heat in a greenhouse. Fourier knew full well that the Earth's atmosphere was not a solid partition, like a glass pane, but he imagined that the atmosphere must in some way block some of the escaping radiation.

In 1859, the British physicist John Tyndall put a finer point on Fourier's idea. Tyndall became interested in the atmosphere through his search for an explanation of the Earth's ice ages; if the composition of the Earth's atmosphere changed, perhaps this could trigger broad climate changes by holding in a greater or lesser amount of the sun's radiation. But what gases were important? To answer this question, Tyndall performed a series of laboratory experiments testing gases known to be in the atmosphere for their transparency to light. Although he found that most gases are virtually transparent, two gases were nearly opaque: most important was $H_2O$, which was contained in the atmosphere as water vapor, and carbon dioxide ($CO_2$), which was released by volcanoes and burning. Tyndall imagined that water vapor was more likely to be implicated in climate change than was $CO_2$, because the concentration of $CO_2$ was relatively low.

In 1896, the Swedish chemist Svante Arrhenius took up the question of the ice ages. Arrhenius added the idea, which had been pursued by a few scientists of the time, that feedback loops might be important. He started with the observation that volcanic activity released $CO_2$, which, according to Tyndall's theory, encouraged warming. Because warmer air held more water vapor, the greenhouse effect was strengthened and this led to even more warming. Similar logic worked in reverse: in a period of low volcanic activity, less $CO_2$ was released to the atmosphere, leading to cooling. Because cooler air held less water vapor, the greenhouse effect was weakened, leading to more cooling.

Arrhenius also did a series of laborious calculations to determine how much the Earth's temperature would change if changes occurred in $CO_2$ levels. His estimate was that if the $CO_2$ levels of the Earth's atmosphere were to double, then the Earth's temperature would go up 5 or 6 degrees Celsius. Arrhenius checked to see if factories and industries could be significant in changing the Earth's climate and found that the $CO_2$ resulting from human activity was about the same as that resulting from natural sources. But he was not at all worried about this scenario. First, his calculations showed that at the then-current rates of $CO_2$ release, it would take a long time, some 1,000 years, for the $CO_2$ concentration to double. Second, he felt that global warming might have good benefits for a country like Sweden, which has a cold climate and a relatively short agricultural season (Fleming, 65).

Other scientists investigated other mechanisms that could change the Earth's climate. Benjamin Franklin noted an effect of volcanoes that worked in opposition to their $CO_2$ release. The enormous aerosols released by volcanoes seemed to provide a protective canopy that shielded the Earth from sunlight and therefore led to cooling. In 1783, volcanic activity in Iceland brought catastrophe for that country's flora and fauna. Franklin speculated that the remnants of the volcanic clouds led to an uncommonly cool summer in Europe.

Another possible cause of climate change and the ice ages was identified as the circulation of water in the Earth's oceans. During the 19th century, scientists identified the basic structure of the Gulf Stream. At the Earth's poles, cooler and saltier water sank below the surface and then flowed along the ocean bottom to the equator. The sinking water acted like a giant pump, pulling up warm water along the surface from the equator. In this way, North America and Northern Europe benefited from steady warm ocean currents. At the start of the 20th century, the American geologist Thomas C. Chamberlin considered the possibility that the circulation might reverse if a delicate balance in ocean salinity were disturbed. Warming would increase the amount of water turned to vapor, which would, consequently, increase the saltiness of the remaining ocean water. If this conversion continued, the saltier and denser water might not proceed north but instead sink to bottom at the equator, thus reversing the global ocean circulation and bringing cold weather to North American (Fleming, 83). (Currently, concern about the Gulf Stream centers more on the possibility of excess fresh water at the poles caused by the melting ice caps.)

Scientists of the late 19th and early 20th centuries also considered celestial mechanics as a cause of global climate change. In the 1870s, the Scottish scientist James Croll imagined that small cyclical changes in the Earth's orbit around the sun triggered the ice ages. He proposed that when the cyclical changes brought a slight decrease in winter sunlight, the accumulation of snow and ice increased. Croll also considered a positive feedback (i.e., a feedback that increases the initial change). As the ice accumulated, Croll hypothesized that an increase of the Earth's overall albedo (or reflectivity) would lead to further cooling (see chapter 1). Unfortunately, Croll's predictions did not match up well with the sequence of four glacial periods then identified by geologists, and therefore his ideas were not widely accepted. (At the time, the glacial periods were referred to as "ice ages," although we now usually reserve this term for the large-scale periods of cold that have since been discovered.)

Croll's work was picked up in the early 20th century by the Serbian engineer Milutin Milankovitch. Milankovitch performed more detailed calculations that showed the Earth's orbit around the sun has three cyclical variations that produce variations in the distribution of solar energy across the Earth's surface. The first two variations are the Earth's orbital eccentricity (how elliptical in shape its orbit is) and the Earth's obliquity (the tilt of its axis of rotation relative to its plane of orbit around the sun). The third variation is particularly important and had a 21,000-year cycle: the Earth's precession around its axis of rotation (like a wobbling top). Similar to Croll, Milankovitch added the idea of positive feedback between a cooling trend and the increase of the Earth's albedo. His calculations of the Earth's precession somewhat agreed with data found by mid century, that the Earth's climate seemed to sway according to cycles of 21,000 years. Although Milankovitch's predictions did not do much better than Croll's at matching a standard sequence of four glacial periods, and raised a number of unanswered questions (such as how a change in the distribution of energy across the Earth's surface resulted in global climate variations), his work was considered to be a promising explanation by at least some climate scientists.

Interest in climate change, the cause of the ice ages, and global warming slowly revived during the first half of the 20th century, as data showed general trends to higher temperatures and later seasonal frosts. During the 1930s, Guy Callendar, an engineer who normally worked on steam engines, took on climate statistics as a hobby. Digging in older publications from the 19th century, Callendar found some evidence that concentrations of $CO_2$ in the Earth's atmosphere had gone up slightly. He therefore returned to Tyndall and Arrhenius's idea that increases in $CO_2$ resulting from human activities might warm the Earth. In 1938, he presented this work at the Royal Meteorological Society. Like Arrhenius, Callendar was optimistic that a slight rise in global temperatures would lengthen the agricultural season. However interesting, Callendar's speculative ideas were based on incomplete data, and they were largely ignored. Other climate scientists agreed that global temperatures did appear to be rising. But the data on $CO_2$ was unconvincing, and the idea that small increases in $CO_2$ could lead to global temperature increases was far from accepted. And so, for another two decades, such ideas languished (Fleming, 113).

## PIECES OF THE PUZZLE

During World War II and the Cold War that followed, climate science received new funding from the federal government, especially from the military, which had a strong interest in predicting and possibly controlling the weather. The American government supported science largely through military channels, most famously through the Office of Naval Research (ONR). For nearly four decades after the end of World War II, climate scientists continued to amass data and develop theories of climate change. Although this work became progressively more detailed and extensive, it proved difficult to put all of the research together to form a coherent picture on global warming. As a result, there was no consensus that human activity was the primary cause of global warming, or even that the trend of global temperatures was upward.

One recipient of ONR's support was Gilbert Plass, a geophysicist at Lockheed Aircraft in California. During the 1950s, he took notice of improved experiments that strengthened the hypothesis that $CO_2$ added to the Earth's atmosphere could block infrared radiation and lead to warming. Plass decided to use digital computers, then in an early stage of development, to calculate changes of temperature resulting from changes in $CO_2$. His trailblazing calculations predicted that $CO_2$ added to the Earth's atmosphere by human activity could raise global temperature by 1.1 degree Celsius each century. Although Plass's model made many simplifying assumptions, rendering its results highly uncertain, it attracted the interest of a number of scientists, including David Keeling, a geochemist at the California Institute of Technology. Keeling constructed new instruments to make precision measurements of $CO_2$ concentrations. After taking measurements in geographical areas that did not generate large amounts of $CO_2$ (away from cities and large farms), Keeling found that the levels of $CO_2$ did not vary much. This suggested that his measurements reflected the average concentrations of the entire globe.

The measurement of $CO_2$ concentrations was encouraged by other, more immediately military, motivations. During the early Cold War, new technology was being developed to detect nuclear bomb tests. The sensitive instruments tested for concentrations of radioactive elements left over by the explosions. Such technology also made it possible to determine carbon levels by tracking "radiocarbon," the carbon isotope $^{14}C$. Because the percentage of $^{14}C$ in the Earth's atmosphere has been reasonably stable throughout the millennia (although at a low level of about 1 atom of $^{14}C$ for every trillion atoms of $^{12}C$), knowing the concentration of $^{14}C$ is a good way to find the total concentration of carbon and therefore $CO_2$.

One of the new experts who measured atmospheric radiocarbon was the geochemist Hans Suess. Suess realized that most of the carbon fuels contained "ancient carbon" that had very little radioactivity (since most of the $^{14}C$ in the fuel had decayed long ago, before it was dug up by humans). Estimates of $CO_2$ levels based on measurements of $^{14}C$ had to be adjusted accordingly. Suess' first results suggested that all the added carbon caused by burning fossil coal and oil was relatively low: only about 1 percent of the carbon in the atmosphere. He therefore proposed that most of the carbon caused by burning was being sucked up by Earth's oceans. This opened up an important question: how fast do the oceans absorb $CO_2$?

At the Scripps Institution of Oceanography (San Diego, California), oceanographer Roger Revelle decided to hire Suess to help him with a study of carbon in the oceans. In 1957, Revelle and Suess wrote a landmark article. While writing the initial drafts, they concluded that the oceans absorbed most of the $CO_2$ added by human activity. Just as they were finishing the article, however, they realize that much of the $CO_2$ absorbed by the ocean would stay at the surface and then return to the atmosphere as water evaporated. In the end, they concluded that $CO_2$ actually takes a long time to be absorbed. Assuming that factories would continue to emit at the 1957 rate (as things turned out, a gross underestimate), they concluded that $CO_2$ levels would increase by 40 percent over 300 years. Just before the article went out for review, Revelle heard of Harrison Brown's work on rising world population. This led the authors to add that, along with the world's population, $CO_2$ levels might rise dramatically. They commented that:

Human Beings are now carrying out a large-scale geophysical experiment of a kind that could not have happened in the past nor be reproduced in the future. Within a few centuries we are returning to the atmosphere and oceans the concentrated organic carbon stored in sedimentary rocks over hundreds of millions of years. This experiment, if adequately documented, may yield a far-reaching insight into the processes determining weather and climate. (Fleming, 125)

Although research on atmospheric concentrations of $CO_2$, such as Revelle and Suess's, had little chance for funding shortly after World War II, fortunes changed as the result of two pressures. First, it was becoming clear that if anything more could be learned about the Earth's geology, meteorology, and climate, such research should be organized at an international level. One example of this is the World Meteorological Organization, which is part of the United Nations. Second, Russia's successful launch of Sputnik 1, on October 4, 1957, galvanized other nations, especially the United States, to put greater funding into scientific research in general and the geosciences and space sciences in particular. These two pressures encouraged a giant international research effort to tabulate geophysical data all over the world, the International Geophysical Year (IGY), which took place from July 1957 to December 1958. Scientists in many countries persuaded their governments to contribute to many different types of research, ranging from Earth's oceans to space.

Revelle decided to seek some IGY funding for a study of the atmosphere and hired David Keeling to do the $CO_2$ measurements. Keeling's perfectionist style led him to construct the most precise (and expensive) instruments for measuring $CO_2$ concentrations in the world. Two measurement stations were established: one in Hawaii and the other in the Antarctic. Before funding ran out in 1960, Revelle and Keeling gathered two full years of data. Even this short data set showed a clear increase in $CO_2$ levels. Revelle received renewed funding in 1961 and there has been continuous monitoring since then. One impressive result is that Keeling's measurements showed the annual variation of $CO_2$ concentrations. Because the Northern Hemisphere has more plants than the Southern Hemisphere, when it experiences summertime, global $CO_2$ concentrations decline (as plants use carbon dioxide through photosynthesis). When it is winter in the Northern Hemisphere, the global $CO_2$ concentration rises slightly. This results in the familiar sawtooth shape of the Keeling curve, which, despite the annual structure, shows a clear trend upward, year by year. This has become one of the most trusted and important data sets in climate science (and perhaps as close as it will ever get to a "smoking gun"). It shows that the oceans are not swallowing up all the $CO_2$ and that a significant amount remains in the atmosphere. Even scientists and policymakers who claim that global warming is not caused by rising $CO_2$ levels must admit that $CO_2$ levels are indeed on the rise.

We have stressed research into understanding $CO_2$ because of its significance to the ultimate issue of global warming; however, many other areas of research were important for the grand puzzle of understanding climate change. One of the important pieces of the puzzle was to extend the climate record beyond the established database of direct measurements going back about

150 years. One of the first people to do this was Cesare Emiliani, a geologist at the University of Chicago. During the 1950s, Emiliani worked with "cores" dug from sea beds. His specialty was the measurement of radioactive isotopes, especially $^{18}O$ found in shells. The amount of $^{18}O$ that the shells absorbed varied in a proportional way with the temperature of the water in which they were formed. This allowed Emiliani to estimate the water temperature at different depths into the sea bed cores, with each depth corresponding to a different time. In this way, Emiliani assembled a record of Earth ocean temperatures for the past 300,000 years (Weart, 45).

The timing of the temperature variations found by Emiliani agreed roughly with Milutin Milankovitch's predictions. Neither Emiliani nor Milankovitch's results, however, agreed with the standard interpretation of geologists, who maintained that the large-scale history of the Earth's climate was more or less exhausted by the history of four big ice ages. Emiliani's response was that the geologists had oversimplified. His reconstruction showed that there had been dozens of cyclic changes, with a period of about 21,000 years. Thus began the study of periods that we now call "glacials" and "interglacials," which occur within the current ice age. A few years after Emiliani's first articles, the scientific community questioned his determination of temperature but agreed that he had established the cyclic rhythm of glacial periods.

By this point in the 1960s, many climate scientists wished to confirm the promising Milankovitch-Emiliani findings. Pursuing this goal led them to a second question: How could climate change relatively suddenly in short, cyclic fluctuations of thousands of years? Some confirming evidence came from work on ice cores. Like sea bed cores, these were long cylindrical samples, removed in sections. Of course, ice cores were taken at the Earth's poles, where ice built up over the centuries and left a record of many thousands of years. One important technique to study ice was developed by the glaciologist Willi Dansgaard around 1954. Like Emiliani, Dansgaard focused on concentrations of $^{18}O$. The more $^{18}O$ that is found in the ice, the warmer the air must have been.

During the International Geophysical Year (1957–1958), researchers working at Camp Century in Greenland used ice cores and Dansgaard's approach to work their way back in the Earth's climate history. Although they could not get a detailed temperature record, they were able to determine clearly the timing of the glacial periods. Their results took time to confirm and were not published until 1969. When they were, the Camp Century results showed that there were a number of overlapping cycles, one as short as 20,000 years and one as long as 100,000 years. The cycles showed rapid heating (over a few thousand years) followed by gradual cooling. Even more exciting, results from Greenland in the Northern Hemisphere and Antarctica in the Southern

Hemisphere were consistent. This strongly suggested that the researchers were looking at truly global changes and not just local variations (Weart, 71).

About four years after the Camp Century work was published, the geologist and paleoclimatologist Nicholas Shackleton studied ice cores with a modified technique that focused on an isotope of potassium. Because the potassium isotope had a longer half-life, Shackleton was able to go back to more distant times, around a million years. His findings agreed with Milankovitch and Emiliani, showing that there were not four main shifts between glacial and interglacial periods but dozens. Work such as Shackleton's on the timing of glaciations also established that orbital variations were of central importance to climate change. Because the orbital variations were so small, however, scientists also became convinced that there must also be natural feedback systems in place to amplify the changes.

Other scientists investigated the variation of the sun itself and speculated that the Earth's mean temperature was higher when sunspot activity was higher. Establishing a measure of past solar activity was pioneered in the early 1960s in a collaboration between scientists Minze Stuiver and Hans Suess. Both focused on the amount of radiocarbon ($^{14}C$) in tree rings. Because $^{14}C$ is generated by cosmic rays interacting with $^{12}C$, they speculated that radiocarbon had an inverse relationship with sunspots. On the one hand, when the sun is less active and has fewer sunspots, there is less solar magnetism. Because the magnetism of sun partially shields the Earth from cosmic rays, less magnetism meant that more cosmic rays got to the Earth and therefore created more $^{14}C$ in tree rings. On the other hand, more solar activity leads to more magnetism and therefore fewer cosmic rays, and therefore less radiocarbon. In 1976, the physicist Jack Eddy did a careful study of the past record of sunspots and found a good correlation between sunspots, radiocarbon, and temperatures. Although some scientists were able to confirm Eddy's correlations, however, others could not. Such work on sunspots was considered promising; however, identifying the trend between sunspots and the Earth's temperature continued to be difficult. In addition, it was unclear how strongly sunspot variation affected the Earth's climate (compared to other climate changes such as greenhouse gasses).

Scientists concerned about trends in global temperatures also investigated aerosols caused by human activity. Walter Roberts, an astrophysicist at the University of Colorado, noted that exhaust from jet airplanes spread out in the skies and became indistinguishable from cirrus clouds. Reid Bryson, a meteorologist at the University of Wisconsin, observed hazes of dust hanging over areas of India, Brazil, and Africa, which were due to fires lit by farmers using slash and burn techniques. The meteorologist Murry Mitchell studied the residue of nuclear bomb tests in the atmosphere. Another concern was smog, which was produced by the sulfur dioxide released to the atmosphere from industrial smokestacks. Sulfur dioxide reacts with water vapor to form particles of sulfuric acid and other sulfates, which hang in atmosphere before they come down to Earth (sometimes after weeks or months) with precipitation. Although such work established that manufactured and natural aerosols were significant, the degree of significance and overall effect (heating or cooling) was a topic of debate for many years.

Another crucial piece of the global warming puzzle concerned land use by human beings. During the 1960s and 1970s, George Woodwell, a botanist with strong environmentalist sympathies, considered the effect of Earth's vegetation on climate. On the one hand, when plants are alive and active, they provided a significant sink for $CO_2$. On the other hand, when plants die and decay, they offered a significant source of additional $CO_2$. Woodwell claimed that deforestation by humans was adding as much $CO_2$ in the atmosphere as was added by the burning of fossil fuels. Oceanographers like Wallace Broecker (see The Discovery of Sudden Climate Change) tended to feel that Woodwell, as a biologist and an environmentalist, was making exaggerated claims. Although the oceans could act as a sink for $CO_2$, the oceanographers felt certain that they could not absorb the levels of $CO_2$ proposed by Woodwell. Eventually, Woodwell accepted that his initial estimates were too high. Nevertheless, by calling attention to deforestation, he had trained attention on an important part of the carbon cycle (Weart, 109).

## THE DEVELOPMENT OF COMPUTER MODELS

As more pieces were added to the puzzle of climate change, scientists were attracted to the idea that there could not be a single primary cause of climate change but instead a tangle of causes and feedbacks. This idea both encouraged and necessitated the use of numerical models to see how all of the proposed climate mechanisms might interact. Some of the earliest computational work showed the difficulties of proceeding by hand calculation. During the 1920s, the English scientist Lewis Fry Richardson sought a calculational system for weather prediction. Richardson followed an approach that is quite similar to modern computer simulations, but on a much smaller scale. He started by dividing up a local territory into grid cells. At each point, his computations keep track of temperature, pressure, humidity, and a number of other quantities. By using known physical laws regarding heat and air flows, Richardson hoped to produce the time evolution of a sample weather pattern. He attempted a calculation of weather over Western Europe for eight–hours. This effort required six weeks of laborious hand calculation but yielded a useless result. Richardson thereafter abandoned his efforts in this area of research (Houghton, 94).

Richardson's work had demonstrated the need for digital computers, which, unfortunately, did not yet exist. The development of computing in the 20th century was greatly encouraged by the effort to develop nuclear weapons. Some of the earliest and most important work on computing was done by the Princeton mathematician John von Neumann. Working at the Manhattan Project (as the American nuclear bomb project was code named) at Los Alamos, New Mexico, von Neumann oversaw the effort to determine the critical mass required for a successful bomb. This was accomplished by simulating many interconnected nuclear reactions (as neutrons from one reaction trigger a further reaction).

After the success of the Manhattan Project, the American military became interested in using computers for the purpose of weather prediction and possibly weather control, and they asked von Neumann to help construct a computer model of Earth's atmosphere. Von Neumann agreed to participate and asked Jule Charney, a meteorologist at the University of Chicago, to head up the project. Charney followed up on Richardson's method of calculation by dividing up the atmosphere into a grid. The computer program (or algorithm) did the mathematical calculations in step-by-step time increments. The results were compared to data painstakingly amassed from different locations and altitudes. Although a great step forward, Charney's efforts at weather prediction were still primitive and offered only regional models. The first global model (a first example of what we now call a general circulation model) was accomplished by Norman Phillips at Princeton University in 1955, concentrating on airflow. Phillips' model gave reasonable results for a typical weather pattern for about 20 days, after which it broke down.

Computer calculations raised a problem that, at first blush, seemed to have a simple answer. Any calculation must start with realistic values of temperature, wind speed, and so on. Phillip's calculations already showed that the smallest changes in initial conditions made big differences in the course of the calculation. The question was this: On the one hand, did this show that the computer model was subject to serious round-off error, giving radically different answers to sets of initial conditions that were virtually identical? Or, on the other hand, did these results indicate something about the real world?

Edward Lorenz, a mathematical meteorologist at MIT, came to the latter conclusion. In keeping with his reputation as one of the founders of chaos theory, Lorenz concluded that a deterministic prediction of weather, in which the initial conditions fully specify the final outcome, was not possible. In the real world of complex, nonlinear systems, the smallest changes in initial conditions led to very different outcomes if one waited long enough. Lorenz summarized this idea in a pithy formula to characterize the behavior of complex systems: "Does the flap of a butterfly's wings in Brazil set off a tornado in Texas?" (Weart, 60, 114).

Although Lorenz asked his amusing question as late as 1979, scientists were convinced of the complexity of global climate by the mid-1960s. The various climate mechanisms that had been elucidated up until that point and their combination in complex computer programs presented a picture to scientists in which the Earth's climate was a fundamentally unstable system. In this vision of climate as a holistic system in which everything is connected to everything else, the idea that tiny changes (in the Milankovitch cycles or in $CO_2$ concentrations) might trigger a large and fundamental change no longer seemed strange as it had 30 years earlier, when Guy Callendar presented this work at the Royal Meteorological Society.

During the late 1960s and early 1970s, the development of models was helped along by two external factors: first, the increase of computing power, encouraged by the development of the microelectronics industry, and second, the amassing of a better database by virtue of balloons, rockets, and weather satellites (such as the Nimbus-3, launched in 1969). The improved database included temperature, pressure, humidity, and wind speed at different altitudes all around the globe.

These tremendous undertakings would be on-going in order to collect the data that might prove or disprove earlier findings. Most well-known, major modeling groups were headed by Yale Mintz at the University of California in Los Angeles and James Hansen at the NASA Goddard Institute for Space Studies in New York City. A third modeling group was based at the U.S. Weather Bureau in Washington, D.C. In 1965, Joseph Smagorinsky and Syukuro Manabe of the Weather Bureau developed a three-dimensional computer model that broke up the Earth's atmosphere into nine levels. Two years later, the team predicted that doubling the level of $CO_2$ in the Earth's atmosphere would raise global temperature by about 2 degrees Celsius. Somewhat later, in 1975, Manabe, now working with Richard Wetherald, developed an improved model that predicted that a doubling of $CO_2$ would result in a warming of 3.5 degrees Celsius. These general circulation models (GCMs) represented a remarkable improvement compared with the hand calculations of Lewis Fry Richardson and began to show scientists how climate processes might be interrelated, but they were still not sophisticated enough to be taken as anything like a "prediction."

One of the main things missing from the early GCMs was the effect of ocean currents. Research on the oceans had identified the large-scale structure of ocean currents such as the Gulf Stream and also many smaller whorls of water circulation. In 1969, Kirk Bryan and Syukuro Manabe collaborated while at Princeton University to combine Manabe's work on the atmosphere with Bryan's work on the oceans, producing the first model to couple the movements of the Earth's atmosphere and oceans. Further work by Manabe and Bryan and others during the 1970s appeared to indicate that the addition of ocean currents somewhat delayed global warming (Weart, 134).

## THE DISCOVERY OF SUDDEN CLIMATE CHANGE

The computer programs of the 1970s were biased toward showing gradual, smooth changes because the assumption before the 1960s had been more or less in line with James Hutton's old idea of uniformitarianism: known geological and climactic processes would operate slowly. By the 1980s, however, climate scientists began to find evidence that the Earth's climate was capable of relatively sudden change.

One example of a scientist who was interested in quick climate change was Wallace Broecker. While a graduate student at the Lamont Geological Observatory at Colombia University in New York City during the 1960s, Broecker became interested in the Milankovitch-Emiliani interpretation of the ice ages. Using radiocarbon dating in sea bed data, Broecker's research group concluded that about 12,000 years ago, at the close of the Younger Dryas cool period, climate went from glacial conditions to relative warmth in as little as 1,000 years. Even Cesare Emiliani criticized this as too short a period of time to be plausible. But Broecker now visualized a new scenario: two climate states that were relatively stable (one glacial and the other interglacial) but that were sometimes kicked from one state to the other in a relatively short period.

Twenty years later, in 1985, after considering further evidence of sudden climate change, Broecker offered a possible explanation in an article titled "Does the ocean-atmosphere system have more than one stable mode of operation?" Broecker (along with two colleagues) answered the article's rhetorical question in the affirmative and, more or less in line with T. C. Chamberlin's speculation 80 years earlier, suggested that the thermohaline circulation could shut down and might be the explanation for the Younger Dryas cool period.

This new concern with sudden climate change was soon embraced by computer modelers. In the second half of the 1980s, modelers like Kirk Bryan and Syukuro Manabe showed that small changes in the key climate parameters, including the $CO_2$ concentration in the atmosphere, could result in relatively quick climate changes. In an article written in 1997 for *Science* magazine, Broecker asked whether the thermohaline circulation was the "Achilles heel of our climate system?" He again answered his question with a "yes" and suggested that "a disturbing characteristic of the Earth's climate system has been revealed, that is, its capability to undergo abrupt switches to very different states of operation" (Linden, 226).

Many other calculations and computer models confirmed that the Earth's climate was sensitive to a delicate balance of physical causes and feedbacks. In the 1960s, many of the models were not yet sophisticated enough to give anything like a prediction of future climate; however, most models had no such ambition. Instead, they were simplified "toy" models, designed to show the general trend of the physical mechanisms being tested. For example, during the early 1960s, Fritz Moeller of the University of Munich tried a series of simple calculations demonstrating, as Arrhenius had suggested so long ago, that water vapor could act as a positive feedback mechanism. Moeller assumed that a doubling of $CO_2$ levels would lead to a rise in global temperature. This initial effect was then amplified greatly by his assumption that the atmosphere's humidity would be proportional to the temperature. Increasing temperature levels therefore led to increasing concentrations of water vapor and therefore more warming. The large increases in temperature (about 10 degrees Celsius) predicted by Moeller with his simple model were later shown to be wrong. Nevertheless, his research demonstrated the plausibility of the positive feedback effects of water vapor.

Equally important research on feedback mechanisms was associated with the hypothesis of global cooling. The idea that there would be a sustained global cooling seemed especially likely during the 1970s after the global temperature record showed a decline for the previous two decades. The conclusion of cooling was never accepted by the mainstream climate science community and was soon rejected, but not before extensive and sensationalist popularization in the media. Nevertheless, the associated research highlighted the possibility of sudden climate change and encouraged research on crucial climate processes. For example, during the late 1960s, the Soviet climatologist Mikhail Budyko contrived a set of equations that kept track of the global radiation budget (see chapter 1) and determined an average global temperature. He found that the overall reflectivity of the Earth's surface, its albedo, was particularly important. Within a certain temperature range, the Earth's climate appeared to be stable. Budyko's equations, however, showed that if the Earth warmed past a critical point, then the system became unstable. Once the ice began to melt at the poles, the albedo went down significantly. After that, Budyko found that the warming-albedo feedback was catastrophic, as the globe continued to heat. The same sort of logic applied to the Earth cooling; beyond a certain point, when ice began to collect at the poles and the Earth's overall albedo began to rise, a runaway cooling trend could result, leading back to the snowball Earth scenario of the first ice ages (Weart, 82).

In 1969, William Sellers did similar but somewhat more detailed calculations at the University of Arizona that confirmed Budyko's findings. Sellers's model kept track of the Earth's land and water temperatures according to 10-degree belts in latitude. The calculation followed four types of energy transfers for each belt: electromagnetic radiation, water vapor flow, air flow, and ocean flow. Like Budyko, Sellers also considered it a distinct possibility that the Earth might be turned into a giant ice ball; however, he also noted the possibility that human activity might kick the delicate balance toward a warming trend.

In 1971, Ichtiaque Rasool and Stephen Schneider published another landmark study. This one used numerical simulation to determine the effect of aerosols in the Earth's atmosphere. Rasool and Schneider found there was a delicate tradeoff between a cooling effect, resulting from the reflection of the sun's radiation, and a warming effect, in which radiation is absorbed by the aerosols. Like Budyko and Sellers, they were particularly concerned about the possibility that a cooling trend might dominate and warned of the possibility of an ice age. Although they checked the possible effect of added $CO_2$ resulting from human activity, they concluded that this did not tip the balance toward warming.

Although these predictions of cataclysmic cooling turned out to be wrong, there is no doubt that the associated research contributed significantly to the modeling of climate processes. A journalistic account of the cooling hypothesis, published in 1976, reflects the fact that, even though there was disagreement about the direction of change, it was agreed that the Earth's climate was unstable and that change could be quicker than had once been believed:

Earth's climate is cooling. This fact seems to contradict theories that say it should be warming. But the prophets of warming are describing real forces that influence climate, and like other scientists they are still learning how these forces interact to produce a balance of heating and cooling on our planet. It may well turn out that the growing instability of Earth's climate is caused by human influences adding both heating and cooling forces to the balance, thereby making it more and more "unnatural" and precarious. (Ponte, 31)

## THE FORMATION OF A CONSENSUS

Before the 1980s, the climate science community had not reached a consensus regarding global warming. The most obvious problem was that there was still uncertainty about temperature trends. Was the Earth undergoing an overall cooling trend, as shown by the data of the last two decades, or a warming trend, as shown by the data earlier in the century? Beyond that, many scientists felt that the climate database about both the current day and the ancient Earth, needed to be fleshed out. Computer models also were deemed to be too primitive to offer forecasts of Earth's climate but were mainly useful to identify the possible interaction of climate mechanisms. Another factor might be that certain scientists maintained a theological or philosophical faith in the idea of a balance of nature, which would render sudden or catastrophic change impossible. In sum, although it seemed possible that human activity might change the global climate, scientists remained unsure about how serious the change might be.

A number of developments during the early 1980s moved the climate community toward consensus. Most important, agreement was reached on temperature trends. Two major groups, at NASA's Goddard Institute for Space Studies (GISS) in the United States, and the Climate Research Unit at the University of East Anglia in England, announced that the global temperature trend was returning to a warming trend, as in the beginning of the century, and away from the temporary cooling trend from the 1940s to the 1970s. In a famous 1981 article published in *Science* magazine, James Hansen and his group at GISS predicted that a clear warming trend would emerge by the end of the century, caused in part by the continuing build up of $CO_2$ in the Earth's atmosphere. Although this bold prediction was not accepted by the climate community right away, each passing year and higher temperatures added weight to the conclusion. In 1986, the group at the University of East Anglia concluded that the three warmest years of the last 134 years had all occurred in the 1980s.

The temperature record of past climate also improved markedly. During the 1980s, the Soviet Union's Vostok Station in Antarctica completed an ice core that yielded an accurate and continuous record stretching back 400,000 years, through four glacial-interglacial cycles. At the same time, the group developed a new method of extracting tiny samples of air trapped at different levels of the ice core and measuring the $CO_2$ levels in the ancient air. The Vostok group made an important finding: $CO_2$ levels were relatively low during the previous four glacial periods and relatively high in between. Although this correlation did not prove causation one way or the other, it did strongly suggest the importance of $CO_2$ in climate change. Because changes in $CO_2$ did not affect one hemisphere in a contrary way to the other, as did the Milankovitch cycles themselves, changes of $CO_2$ suggested a promising positive feedback mechanism whereby small changes in the Earth's orbit touched off larger changes in global climate (Weart, 126).

Despite these tremendous leaps in scientific understanding, climate science by the 1970s continued to lack the coordination of a large, international group of scientists. National and international political pressures also pointed to the need for coordinated research. As discussed in chapters 4 and 5, a number of environmental issues surfaced during the late 1970s and 1980s (including air pollution, acid rain, and ozone depletion), which gained the attention of the general public and policymakers. After the heat waves and droughts of the spring and summer of 1988, and Jim Hansen's warnings about rising temperatures in his testimony to Congress, the stage was set for global warming to be acknowledged as a problem facing all of humanity and requiring further investigation.

Such scientific and political pressures called out for the formation of an international body of scientists to address the problem. In 1988, The World Meteorological Organization and the United Nations did just this in forming the Intergovernmental Panel on Climate Change (IPCC). The main purpose of the panel was to produce periodic reports summarizing the best knowledge on global warming. Because many scientists associated with the IPCC had connections with various scientific agencies and national laboratories in their home countries, the panel can be thought of as a sort of hybrid organization, both scientific and political.

In 1990, the IPCC released its first report, the product of work by about 170 scientists working in about 12 different workshops. After each workshop wrote its own report on each specific topic, the reports were sent to scientists outside the IPCC for criticism and comments. After revision, they were then synthesized into a single IPCC report. In so doing, the material was organized into three major sections. The first concerned the scientific basis of the work, including everything from climate data to computer modeling. The second centered on an assessment of the impacts of climate trends on human society, translating from the language of greenhouse gas concentrations, temperature changes, and model results to more specific forecasts involving changes in water supply, ecosystems, and coastlines. Finally, the third report suggested possible mitigation or adaptation responses. The purpose of the third group was "in broad sense technical not political," and "to lay out as fully and fairly as possible a set of response policy options and the factual basis for those options" (IPCC, 1991).

Because the IPCC was faced with not only a difficult scientific problem but also a tricky political position, it stated its conclusions cautiously and in qualified language. Nevertheless, the report did reach some important conclusions. For one thing, the scientists agreed that the natural greenhouse effect was on the rise. They found that the average global surface air temperature had increased by 0.3 to 0.6 degrees Celsius over the previous 100 years. Over the next 100 years, the panel estimated that the Earth would warm by something in the range of 1.5 to 4.5 degrees Celsius. With the ensuing partial melting of Earth's ice caps (especially in the north) and thermal expansion of the water itself as a result of increasing ocean temperatures, the sea level would rise at an average rate of roughly 6 cm per decade. The report also concluded that although some of the warming might be due to natural processes, a portion was probably due to human activity, which was increasing the atmospheric concentrations of the greenhouse gases.

In all, the first IPCC report sent a mild warning. But it also sent a signal that an international scientific consensus on global warming was forming. This proved to be the case in the further reports issued by the IPCC in 1995, 2001, and 2007, to which we turn in chapter 5. At the same time that the scientific consensus on global warming was solidifying, so was the issue's political significance. Political pressure led to a number of initiatives, the most important being the United Nations Framework Convention on Climate Change (UNFCCC) endorsed at the Earth Summit in Rio de Janeiro in 1992, and the Kyoto Protocol, adopted at the UNFCCC's third Conference of the Parties in Kyoto, Japan in 1997. These are also the subject of chapter 5. For the moment, we close the present chapter by reviewing some of the scientific objections that were raised against the young consensus.

## SKEPTICAL CHALLENGES

Although the science and politics of climate science were probably always intertwined, this connection increased as the issue took the political stage during the 1990s. During this time, scientists who were skeptical of global warming gained opportunities to work with various conservative think tanks such as the George C. Marshall Institute (founded in 1984) and the Global Climate Coalition (1989), both of which were funded by companies from the fossil fuel and automobile industries. The overriding ideological concern of such groups—to resist government regulation, which they believed would hurt the U.S. economy—is discussed further in chapter 4.

A first series of scientific challenges denied that global warming was even occurring; a second agreed that temperature trends were indeed rising but that human activity had little or nothing to do with it. The validity and seriousness of these challenges varied just as widely as the topics. At one end of the spectrum, many skeptics correctly identified weak points in the global warming consensus and, in fact, many of these issues were subjects of ongoing research. At the other end of the spectrum were objections that had already been raised and answered a number of times before. Past a certain point, such challenges were not given serious attention by the mainstream climate science community or accepted for publication in peer-reviewed journals. This fact was taken by the more obdurate skeptics as an indication that global warming researchers formed a closed society that did not wish to endanger its funding and position of prestige.

During the 1990s, the mainstream community acknowledged that there were many difficulties concerning, for example, the temperature record, the computer models, the importance of solar activity, and the relative importance of positive and negative feedback mechanisms. As the remainder of this section shows, the climate community judged that these difficulties were not great enough to reject the consensus. But, as shown in chapter 5, there was yet a third type of skeptical argument (popularized by Bjørn Lomborg) that accepted that warming was occurring and that human activity was a significant cause, but warned that acting on the problem was prohibitively expensive.

One recurring criticism from the global warming skeptics of the 1990s was that the temperature record was flawed. A common objection was that the temperature increases were spurious because so many of the measurements were taken near urban areas. Over the years, cities and roads grew, and the modification of the land surface led to higher local temperatures, giving a false impression of warming. Another claim was that global temperatures appeared to be rising but that these merely reflected an unimportant statistical fluctuation, which would soon return to a cooling trend, just as had happened between the 1940s and 1970s. Such objections, however, had already been considered by the climate science community and had been answered to their satisfaction. The temperature records had been corrected for the heat-island effect and had been subjected to repeated statistical analysis. The IPCC reports reflected a broad consensus on both issues.

Some skeptics accepted that the Earth's average surface temperature was indeed rising but that this was not caused by human activity. Instead, the true cause of the rising temperatures, they claimed, was increased solar activity. Indeed, the computer models of the time suggested that slight increases in solar activity could possibly trigger rising global temperatures (helped along by other feedback mechanisms). Research suggested that there was a correlation between sunspot activity and global temperatures during the past 1,000 years or so. It also suggested that the warming trends of the late 19th and early 20th centuries might be explained, in part, by increased solar activity (especially seeing as how the greenhouse effect could not have been significantly enhanced at that point). The case for solar activity, however, broke down for recent decades. All available evidence pointed to the fact that solar activity was no greater in the late 20th century than it was at the start. Furthermore, using the solar-activity argument to reject the hypothesis of the enhanced greenhouse effect proved to be a double-edged sword. Demonstrating that slight changes in solar activity had in the past led to significant changes of

Earth's temperature lent credence to the idea that other slight changes might lead to significant climate change. Meanwhile, other data and model results pointed to the possibility that slight changes in $CO_2$ levels could result in a significant radiative forcing (or perturbation to the Earth's energy budget). Such considerations further cemented the consensus view that the warming trend of the late 20th century was largely due to rising $CO_2$ levels, despite repeated objections from the more stubborn skeptics (Weart, 167).

Another common set of criticisms aimed at the computer models. The different GCMs gave somewhat different results and sometimes contradicted the known climate record. One specific contradiction concerned data on the climate of the previous glacial period (before the beginning of the Holocene epoch). Data were gathered during the mid-1970s by an oceanography project called "Climate Long Range Investigation, Mapping, and Prediction" (CLIMAP), pursued by a consortium of universities. CLIMAP concluded that the temperature record contradicted the results of the best computer models. Rather than being cold, as the models had predicted, the actual ocean temperatures of the previous glacial period appeared to be not much colder than they were in the current day. This anomaly persisted until the late 1990s, when further measures of past climate suggested that it was not the computer models that were faulty but rather how the CLIMAP oceanographers had analyzed their data.

This is a good place to note a general confusion that surrounded the computer models of the 1990s, one that persists to the present day. Many observers objected to the fact that the climate models did not provide reproducible "predictions" of climate change. The fact is, that although the word *predictions* is often used to describe the results of computer models, the word should not be used uncritically. The computer models unavoidably contain many uncertainties involving the empirical data, the theory of climate processes, and unknowns about future human activity. Despite the sensitivity of the simulations to such input, the climate science community has shown that the models can reproduce past climate and give reliable indications of future trends. In addressing this issue, the IPCC uses terms like *forecasts* and *scenarios* to describe model results instead of *predictions*. Each individual scenario has uncertainties associated with it and makes different assumptions, especially concerning future human social and economic conditions (Houghton, 138).

Further scientific challenges during the 1990s had to do with greenhouse gases. One common objection was that if levels of $CO_2$ were indeed increasing, then it might actually be a good thing, as rising $CO_2$ in the atmosphere would benefit crop growth and agricultural production (since plants would have more available $CO_2$ for photosynthesis) and that this was a significant negative feedback counterbalancing global warming trends. Although it was true that the Earth's plants to some degree sopped up the extra $CO_2$, this not only benefited desirable plants but also undesirable ones (aka weeds). And, although it proved true that the $CO_2$ fertilization effect led to a significant negative feedback, this was counteracted by a number of other positive feedbacks due to the Earth's biomass (such as the increase of brushfires and forest fires, which served to reduce photosynthesis).

Another objection was raised by meteorologist Richard Lindzen, who questioned the positive feedback loop between warming and water vapor. Lindzen shared the bias of a number of other scientists and policymakers, that nature must necessarily favor balance. He sought to demonstrate that the warming-water vapor interaction led to negative feedback and therefore acted to balance and stabilize the Earth's climate. But, once again, this possibility had been considered and rejected by the climate science community. Despite some uncertainty about the warming-water vapor relationship, climate scientists agreed that the two were connected predominately by a positive feedback (Weart, 169).

As mentioned previously, there also were significant uncertainties regarding aerosols in the Earth's atmosphere. Although it was relatively clear that natural aerosols (from volcano releases, forest fires, and dust blown off the land) served to lower regional temperatures, the overall effect of manufactured aerosols was not certain. For one thing, it was not clear if manufactured pollution was more or less significant than volcanic activity. For another thing, there was debate about whether manufactured aerosols led to cooling or warming. In 1976, Bert Bolin and Robert Charlson suggested that sulfate aerosols could cause cooling by blocking sunlight. But one year later, Sean Twomey countered that such aerosols could cause warming by absorbing energy and then reradiating it back to Earth (adding a greenhouse effect of their own). During the ensuing decades, scientists judged that sulfate aerosol levels resulting from human activity were of greater importance than those from volcanoes (Weart, 130). They also came to favor the conclusion that manufactured aerosols produced a small cooling effect, or "global dimming" (Burroughs, 203).

In addition to answering or making headway with a large number of ongoing debates and outright challenges, the climate science community gained new and exciting scientific results. Some of the most important confirmed that global temperature changes could be larger and quicker than previously accepted. For one thing, still better ice cores became available, this time from two teams in Greenland, one American and the other European. The results of both studies were in good agreement. Alongside relatively slow climate changes, the ice-core scientists also identified a number of surprisingly quick changes. Some examples were associated with the Younger Dryas period, the relatively short cooling period before the start of the Holocene epoch (see chapter 1). The ice cores showed that temperatures changed by as much as 7 degrees Celsius in a matter of decades (Weart, 174).

The ice core results also raised an important question often posed by skeptics of the hypothesis of anthropogenic global warming. Although the ice cores showed a tight correlation between temperature and $CO_2$ levels, the causation was not clear. Whereas one might expect $CO_2$ levels to lead to temperature increases, the data often suggested that $CO_2$ lags the temperature increases. Although, to this day, there is still uncertainty about the timing of each and every variation in ice core data, these results are not in contradiction with plausible physical mechanisms. In 1990, James Hanson collaborated with the Vostok scientists to conclude that there are a number of possible ways that feedbacks might combine to bring about significant warming. In one scenario, a small initial warming trend might be due to changes in the Milankovitch cycles; however, a number of feedbacks could then encourage increases of atmospheric $CO_2$ (for example, because the solubility in the oceans generally decreases with rising temperature and because higher temperatures and fires reduce the biosphere and thus photosynthesis). The $CO_2$ feedbacks act to magnify greatly the original small change brought on by the tiny orbital changes (Bowen, 245).

As if mirroring the sudden changes in the ancient temperature record, temperature measurements of the last decade suggested that the recent warming was greater than anything for a thousand years. In 1998, the climate scientists Michael E. Mann, Raymond S. Bradley, and Malcolm K. Hughes (MBH) published a trailblazing article that summarized the results of proxy temperature measurements (especially tree rings) for the Northern Hemisphere. The authors encapsulated their results in a graph showing the temperature increase during the last thousand years. This became widely known as the "hockey stick" graph because of its relatively flat shape before the early 20th century and then the sudden rise of the temperature increase, reaching about a half a degree. We return to controversies regarding the MBH findings in chapter 4.

New computer models, in part motivated by the new empirical information, also showed that sudden climate change was possible and could be explained by known physical mechanisms. In addition, the new codes were now able to show local variations of climate along with the trends of the entire globe. Estimates of the rate of climate change had changed dramatically. Whereas in the 1960s, climate changes were thought to be gradual, on the order of 10,000s of years, this number declined steadily. By the 1970s, significant climate change was seen to be possible in 1,000s of years, in the 1980s, 100s of years, and finally in the 1990s, in just 10s of years! Of course, such a finding greatly supported the idea that human activities (which operate on the scale of generations) could be an important determinate in global warming.

Many of the skeptical challenges of the 1990s were significant and required hard work by the climate scientists to build and maintain their fragile consensus. In many ways, the work of researchers at centers like the Goddard Institute in New York City and the Hadley Centre in Exeter, as well as international bodies like the IPCC, represented a culmination of more than 100 years of scientific research. What had started as an interesting speculation at the turn of the 20th century (with the work of Svante Arrhenius) had now became a well-developed research field involving thousands of scientists worldwide. It also presented itself as one of the most complex and worrisome problems confronting human civilization. As such, it became a political football, both in the United States and internationally, a story that we address in chapters 4 and 5. In chapter 3, we back up and answer the question: "If the hypothesis of anthropogenic global warming is correct, then how did human beings get themselves in a position where they risked changing the Earth's very climate?"

# 3

# How Did We Get Here?

There was no particular "eureka!" moment that ensured humans learned to burn fossil fuels in order to release stored energy. Instead there was a steady stream of innovations teamed with a consistently growing willingness to derive work and labor from stored reservoirs of energy instead of from actual bodies—those of animals and humans. Overall, the greatest single change brought by this shift, which is normally referred to as the Industrial Revolution, was the release of human accomplishment from the limits of the caps on human and animal energy. Humans living throughout the world developed new methods for living; for this reason, although our emphasis in this chapter is American industrialization, the chronology first traces these innovations in other parts of the world. As some of these patterns became more prevalent, they reshaped the expectations for all humans as a species.

Plank by plank, this chapter depicts the shift in the human condition in which energy emerges as a most crucial component. By the close of the 20th century, the standard paradigm of energy use will derive from the burning of hydrocarbons. This way of harvesting power will be so accepted and normal that, at that juncture in the 20th century, other methods of obtaining energy—even if they have been used for thousands of years—will become classified as "alternatives." Burning fossil fuels was the new normal. Today, as we return to some of these alternative methods, we place the era of industrialization within a critical portion of historical context: the expansion of the last 250 years that was made possible by the burning of fossil fuels has contributed to—if not entirely stirred—climate change in the 21st century.

The context for this chapter is the scientific consensus discussed in chapter 2: that the details of this shift in human living carried consequences for Earth's climate. The Industrial Revolution included revolutionary changes for many aspects of human life—many of which are remarkably positive, such as life expectancy and expanding opportunity for economic success. As chapter 2 demonstrates, however, we have begun to comprehend that this era of growth and expansion for many humans did not come without a cost.

## REORGANIZING HUMAN LIFE

Energy has always been a part of human life. Humans began the management and harvest of Earth's energy with their existence as hunter-gatherers. As their ability to manage and to manipulate the surrounding natural systems matured, humans passed through what historians and archaeologists refer to as the Agricultural Revolution. This shift in human life occurred at different moments throughout the globe within the last few thousand years. Adapting to climatic variations, humans in different regions took control of the natural cycles of energy—primarily of the sun and photosynthesis—and learned to condition their behavior, resulting in a relatively consistent supply of food. Once food management had allowed humans to become more sedentary, they only had to make a slight adaptation to their living patterns in order to exploit and to develop the practices that we refer to as early industry.

The use of water and wind power grew closely with agricultural undertakings. These power sources have been used in milling for centuries. For instance, mills powered by waterwheels were used to grind grain from at least the first century. The Domesday Book survey of 1086, the primary written source of its era, lists 5,624 mills in the south and east of England. Similar technology could be found throughout Europe and elsewhere, applied to milling or other tasks, including pressing oil or even making wire. Most often, each of these industrial establishments was an entirely local, limited endeavor. A few exceptions also grew well beyond the typical village center. For instance, near Arles in France, Roman builders used 16 wheels in a single mill. In each case, however, the energy was harvested and applied to a specific activity; it did not necessarily alter the way most humans lived their daily lives.

The organization and adaptation that historians refer to as the Industrial Revolution came much later, when technical innovations came to form dominant patterns in human life. First, however, these energy sources were used in areas where human and capital concentration made it more possible. Some of the earliest milling technology arrived in England through use in religious communities including monasteries.

Monasteries at this time were self-sufficient religious communities producing their own food and other goods. They were often referred to as estates, and they seem to resemble diversified plantations. One of these enterprises had monks turning wool into various forms of cloth by wetting and pounding it. The name of this process was "fulling." The cloth would be placed in a trough filled with the fulling liquor and then would be walked on with bare feet to thicken it. This process was revolutionized when the Cistercians at Quarr Abbey set up a mill that would full the wool by using water power. Although this was not the first fulling mill in England, historians credit it with initiating the enterprise on the Isle of Wight, which became world-renowned for its kerseys, a coarse cloth made on a narrow loom. With access to a waterwheel, the monks created a series of large wooden mallets that would pound on the fabric while it was in the liquid, making the cleaning and thickening process much more rigorous and even. Perfecting these methods allowed merchants to prepare for important shifts in European history that were about to arrive.

The wars of the Renaissance and Reformation eras proved to be a great boon for merchants and manufacturers supplying armed forces. Many of these new industries and systems of transportation would ultimately be put to peacetime uses as well. By most modern measures, however, the manufacturing taking place from the 1300s to the 1500s was of a very limited scale. Between 1500 and 1750, changes in manufacturing continued, but they would not accelerate remarkably until after 1750.

During this early era of manufacturing, most enterprises garnered energy from passive means including rivers and the wind. Historians often refer to this era as "protoindustrialization," when cottage industries applied new devices and methods in concentrated efforts. Each source of power proved extremely limited in energy and reliability. Of course, this meant that manufacturing also could not be reliable and could expand only to a limited level.

The manufacturing that did develop was most often based on technologies that European merchants brought from other regions, particularly from Asia. For instance, Europeans perfected the art of making porcelain imitations of Chinese crafts. And from India, Europeans imported methods for manufacturing silk and textiles. While perfecting these technologies, European business leaders also linked specialized, craft production into larger scale systems that also placed small-batch into the class of manufacturing.

The basis for this system of manufacturing was improved energy resources. Ultimately, the outcome was a large-scale shift in economic and social patterns in Europe that culminated with the formation of an entirely new organization to society. These living patterns ultimately led up to and fostered the Industrial Revolution.

## URBANIZATION AND MANUFACTURING

Technological innovations carry with them social and cultural implications of great import. For instance, industry brought new importance for people to settle and live in clustered communities. Even limited energy development, such as the water wheels, spurred urbanization in human history. Changes in manufacturing from 1300 to 1650 brought with it major alterations to the economic organizations of European society, as well as the availability of goods and services. In addition, however, patterns such as urbanization helped to foster other factors that helped a singular innovation move into the realm of industrial development.

For instance, the growth of urban areas brought profound changes in banking and in the technology that supported manufacturing. A class of big businessmen arose, and in connection with it an urban working class, often referred to as the proletariat. Many related sectors of society had to change after these economic developments. For instance, people of influence rethought basic legal institutions and even ideas of property. For years complex sets of obligations had limited property holding; now, city dwellers could hold property in their own right. This liberation and flexibility of capital were critical to later economic developments. The ability of business holders to collect and improve value in property became a crucial component of later development and expansion. Such basic freedoms led directly to the development of towns as trade, banking, and manufacturing acquired a new scale. As an increasing number of persons achieved the legal status of free men, they were able to improve their economic condition, become land owners, and even establish their own enterprises. New lands were needed for this new group, and many societies worked to convert uninhabitable forest or swamp for cultivation. A new order took shape on the landscape. Even though it was not yet industrial, its organization around agricultural production proved a crucial stepping stone toward industry.

These early industries rapidly made flexibility a valuable commodity and increased the potential of undertakings that did not rely on geographical features such as wind, tidal flow, and river power. For instance, early industries quickly began to have an impact on Europe's supply of wood. During this early period of industry, Western Europe's forests largely disappeared as they served as the source of raw material for shipbuilding and metallurgy. This shortage led English ironmasters, however, to use a new source of energy that would greatly multiply the scale and scope of industrial potential. The English use of coal and more specifically of coke revolutionized the scale and scope of the manufacturing that followed throughout the world.

Although Western Europe had abundant supplies of ordinary coal, it had proven useless for smelting ore. Its chemical impurities, such as phosphorus, prohibited its generation of strong iron. For this reason, smelting was fired with charcoal, which was made from wood.

Western Europe's lack of wood made it lag behind other regions during these decades. In approximately 1709, however, Abraham Darby discovered that he could purify coal by partly burning it. The resulting coke could then be used as a smelting fuel for making iron. Darby released this knowledge for public use in 1750. This process proved to be a launching point for the reliance on fossil fuels that would power the Industrial Revolution.

## THE INTELLECTUAL UNDERPINNINGS OF THE MACHINE

The period of early urbanization and manufacturing, which lasted from 1500 to 1750, can best be described as encompassing great technological developments but no genuine "revolution" of industrial expansion. In an era in which scientific and technological innovations were frowned upon and energies and monetary support were focused on exploring the globe, it is relatively remarkable that any developments occurred at all. Simply, society of the Reformation was not conducive to new technological developments. The pressure to conform in this era slowed technological change and kept the implications of energy development fairly focused.

Nevertheless, social changes did occur that bore a significant impact on later uses of technology. Industry began to move outside of cities. The nation-states that began to develop slowly became somewhat supportive of select technologies, especially technologies and machines that might be used in battle, including designing fortifications, casting cannons, and improving naval fighting ships.

More important to most members of society, during the 18th century the manufacture of cotton ushered in wholesale changes in the way humans did work. A series of inventions in England combined to create a way of work that became known as the factory system. In short, this system brought machines into the lives of most factory workers.

Over the course of the century, various branches of industry were stimulated by similar advances, and all of these together, in a process of mutual reinforcement, launched an entire era grown, at least partly, on the back of technological gains. The age would be organized around the substitution of machines for human skill and effort. Heat made from inanimate objects took over for animals and human muscle. Furthermore, this shift enhanced the amount—the scale and scope—of the work that could be undertaken.

After 1750, of course, the steam engine and related developments generated a bone fide Industrial Revolution. As Joel Mokyr has written, "if European technology had stopped dead in its tracks—as Islam's had done around 1200, China's by 1450, and Japan's by 1600—a global equilibrium would have settled in that would have left the status quo intact" (1990, 52–3). Instead, of course, over the next two centuries, human life changed more than it had in its previous 7,000 years. At the root of this change lay machines and an entrepreneurial society committed to applying new technologies to everyday life, each one relying on new, flexible, and expandable sources of energy.

## ENERGY SOURCES FUEL THE INDUSTRIAL TRANSITION

What historians of technology refer to as the great transition is not necessarily the emergence of the Industrial Revolution in the mid-1700s. To reach that revolution, a "great transition" was necessary in intellectual thought and in the availability of energy resources. Biomass fuels such as wood and charcoal had been in use for centuries, but they could not support an entirely new infrastructural system of machines. By contrast, coal emerged as a prime mover during the 1600s and did exactly that.

After England experienced serious shortages of wood in the 1500s, domestic coal extraction became the obvious alternative. Most of the existing coal fields in England were opened between 1540 and 1640. By 1650, the annual coal output exceeded 2 million tons. It would rise to 10 million tons by the end of the 1700s. Mining technology, of course, needed to be developed quickly to provide the fuel to power this new era. In the new energy resource of coal, industrialists found potential power that far exceeded any sources then in use. Thus new industrial capabilities became possible. Primary among these was the steam engine.

The basic idea of the steam engine grew from the exploration of some of the revolutionary intellects of this new era in human history. Scientific minds were becoming freer to openly explore innovations that might significantly alter human life. For instance, the idea of the piston, which was the basis of the engine, came about only after the realization of the existence of Earth's atmosphere. Although other societies had thought about the concept of an atmosphere and pressure holding things to Earth, it was Europeans who began to contemplate the possibilities of replicating this effect in miniature.

In the mid-1600s, English engineers began contemplating a machine that used condensation to create a repeating vacuum to yield a source of power.

The first model of such a device is attributed to Denis Papin who, in 1691, created a prototype piston that was moved within a cylinder using steam. This device remained unreliable for use, however, because the temperature could not be controlled.

In 1712, Thomas Newcomen used atmospheric pressure in a machine that he alternatively heated and cooled in order to create the condensation pressure necessary to generate force. Also, Newcomen's engine was fairly simple to replicate by English craftsmen. Used to pump out wells and for other suction purposes, the Newcomen engine spread to Belgium, France, Germany, Spain, Hungary, and Sweden by 1730. Although it lacked efficiency and could not generate large-scale power, the Newcomen engine was a vision of the future. It marked the first economically viable machine to transfer thermal energy into kinetic energy. This concept, powered by a variety of energy sources, was the flexible prime mover that would lead the Industrial Revolution.

The need for energy sources and the trade networks forming in the Atlantic provided another portion of the raw material to spread industry. Linked by ships, European powers sought necessary resources in other regions. Soon, this led the mercantilist nations to establish colonies. In North America, settlement grew from agriculture; however, as the United States developed, it emphasized industries using technologies perfected in Europe and new ones that blazed important new paths. The key connecting each undertaking was that energy was the necessary raw material to develop the young nation. Known as the "grand experiment," the United States functioned almost as a laboratory for expansively applying industrial ways to developing an entire nation.

## COAL SPREADS A NEW INDUSTRIAL ERA

America of the early 1800s still relied on energy technologies that would be considered sustainable and alternative to fossil fuels. The transition, however, had begun as industrialists expanded the use of charcoal, which created an infrastructure that could be expanded to include additional energy sources. Some of these resources, however, were complicated to harvest and manage. Their acquisition demanded entirely separate technological innovations, as well as shifts in the accepted patterns of human life.

In the early 1800s, timber or charcoal (made from wood) filled most Americans' heating and energy production needs. This changed rather suddenly during the War of 1812, which pitted the United States against Great Britain in a conflict over trade and ended in stalemate in 1815. The root of the conflict was the rights of American sailors who were being impressed to serve in the British Navy. The major military initiative of Britain during the war was more related to trade; the British blockade of ports such as Philadelphia nearly crumbled the economy of the young republic.

The blockades of the War of 1812 became instrumental in moving the United States more swiftly toward its industrial future. Depleting fuel wood supplies combined with the British blockade to create domestic interest in using anthracite or hard coal, particularly around Philadelphia. Historian Martin Melosi writes, "When war broke out . . . [Philadelphia] faced a critical fuel shortage. Residents in the anthracite region of northeastern Pennsylvania had used local hard coal before the war, but Philadelphia depended on bituminous coal from Virginia and Great Britain." Coal prices soared by more than 200 percent by April 1813. Philadelphia's artisans and craftsmen responded by establishing the Mutual Assistance Coal Company to seek other sources. Anthracite soon arrived from the Wilkes Barre, Pennsylvania area. After the war, industrial use of hard coal continued to increase slowly until 1830. Between 1830 and 1850, the use of anthracite coal increased by 1,000 percent (Melosi, 1985, 130).

This massive increase in the use of anthracite demonstrates how the Industrial Revolution, in a larger sense, represented a transitional period, with animate, muscular energy being almost entirely replaced by inanimate, hydraulic-based energy. Steam engines converted coal's energy into mechanical motion but still remained very limited in their application. Building on the early work of Newcomen and others, James Watt created an engine in 1769 that did not require cooling, which allowed the use of steam to spread. Ultimately, then, during this same era water-powered milling was replaced by inanimate, fossil-fuel-based energy in the form of steam power.

As the Industrial Revolution swept from Europe into other parts of the world in the early to mid-1800s, the nations most susceptible to its influence were rich in raw materials and committed to the individual freedom of economic development. In these considerations, the United States led the world. Thanks to the American interest in free enterprise and the astounding supplies of raw materials, including coal and later petroleum, the United States became the industrial leader of the world by the early 1900s, after only four or five decades fully committed to industrialization. Economic prosperity, massive fortunes for a few, and employment for nearly everyone who wanted to work were a few of the outcomes of American industry. Another outcome, however, was the environmental degradation that resulted from the intense use of the natural resources exerted by industrialization.

In the industrial era that stretched from 1850–1960, many industrialists were willing to create long-term environmental problems and messes in the interest of short-term gain. Some of these gains came in the form of unparalleled personal fortunes. Other benefits included long-standing economic development for communities and regions around the United States. This economic strategy took shape on the back of the harvest, manipulation, and exploitation of natural resources. This ethic of extraction was felt to some degree in any industrial community, but possibly it was most pronounced in mining areas, particularly those areas mining for energy resources such as coal and petroleum (Black, 2000).

As American society committed to a primary course of development that was powered by fossil fuels, much of the evidence of extraction and production was viewed as symbols of progress. Few checks and balances existed to demand care and conservation. In the 19th century, the environmental consequences of mining for these hydrocarbon resources buried deep in the Earth was of little concern. Most often, industries were viewed almost solely for the economic development that they made possible. Smoke, pollution, and the effluent of industrial undertaking marked progressive symbols of economic expansion.

## THE HYDROCARBON PAST PROVIDES ENERGY FOR THE FUTURE

In terms of energy production, the Industrial Revolution marked the moment when humans turned to the flexibility and concentrated energy within minerals such as coal. Created from 100- to 400-million-year old plant remains deposited when huge swampy forests covered parts of the Earth, coal had been mined by humans since the era of ancient Rome. Formed over millennia, coal cannot replenish itself, and hence it is a non-renewable energy source. The energy we extract from coal and petroleum today is essentially the energy absorbed by plants from the sun millions of years ago. This transfer process is known as photosynthesis.

Once plants die, they release the energy that they have absorbed from the sun. When coal forms, this process of decay is interrupted, and the plants' energy is not lost. The stored solar energy can be retained for millions of years if conditions are just right. Geologists believe these perfect conditions have occurred most often when dead plant matter has collected in swampy water. Such a collection can slowly form a thick layer of dead plants decaying at the swamp's bottom. As the surrounding climate changes over time, additional water and dirt can halt the decaying process. As the upper layers of water and dirt press down on the lower layers of plant matter, they create heat and pressure. Spurring chemical changes, these catalysts press out the plants' oxygen and leave behind rich hydrocarbon deposits. This material hardens and forms coal. Occurring in seams that can range from a fraction of an inch to hundreds of feet deep, coal winds itself into Earth's crust (Secondary Energy Infobook). For instance, the well-known Pittsburgh seam is approximately seven feet thick and is thought to represent 2,000 years of rapid plant growth. The potential energy in such a seam is remarkable. Experts speculate that this seam contains about 14,000 tons of coal, which would fill the electric needs of 4,500 American homes for one year.

Although petroleum would become a vital cog in portions of the industrial era, coal was the prime mover. Of course, coal deposits are scattered throughout the globe; however, northeastern Pennsylvania has a unique 500-square-mile coal region. When coal was being formed millions of years ago, northeastern Pennsylvania endured a violent upheaval, referred to by geologists as the Appalachian Revolution, that accelerated the process. Geologists speculated that the mountains literally folded over, exerting extra pressure on the subterranean resources. In northeastern Pennsylvania, this process created a supply of coal that was purer, harder, and of higher carbon content than any other variety. Named first with the adjective *hard,* this coal eventually became known as anthracite. Geologists estimate that 95 percent of the supply of this hard coal in the Western Hemisphere comes from this single portion of northeastern Pennsylvania.

This supply defined life in the state during the late 1800s. Thousands of families from many different ethnic backgrounds moved to mining towns to support themselves by laboring after coal. In other areas, mills and factories were built that relied on the coal as a power source. In between, the railroad employed thousands of workers to carry coal and raw materials to the mills and finished products away from them.

Coal would alter every American's life through the work it made possible. Although coal was found in a few Mid-Atlantic states, Pennsylvania possessed the most significant supplies and, therefore, became "ground zero" for the ways in which coal culture would influence the nature of work and workers' lives in the United States. The coal communities that grew during the anthracite era reflected the severely hierarchical organization that defined labor in the coalfields. At its top, an elite class of mine owners and operators often lived in splendid mansions, while the immigrant laborers at the bottom lived in overcrowded, company-owned "patch towns." The class disparity was perpetuated by the steady change in ethnic laboring groups that occurred as waves of new immigrants arrived in the coalfields of Appalachia. The original miners from Germany and Wales were soon followed by the Irish and later the Italians, Poles, and Lithuanians.

Out of these difficult and divided living conditions, diverse ethnic groups ultimately created vibrant enclaves. In each patch town, the various ethnic groups built churches, formed clubs, and helped others from their nation of origin to get a start in the fields. A culture evolved around these coal towns that often remained when the coal patches were gone.

In Britain and the United States, the use of coal also came to define communities. By the mid-1800s, many British factory towns were demarcated by tall chimneys, shaped from brick, stone, iron, and concrete. Although these "Cleopatra's Needles" served as symbolic landmarks of industrial development, historian Gale Christianson also argues that they were one of the first functional responses to awareness of the unpleasant outputs of burning coal. In *Greenhouse,* Christianson cites Robert M. Bancroft's 1885 tract *Tall Chimney Construction* for an explanation of why tall chimneys were needed: "Firstly, to create the necessary draught for the combustion of fuel; secondly, to convey the noxious gases to such a height that they shall be so intermingled with the atmosphere as not to be injurious to the health" (Christianson, 55). By the 1840s, brick chimneys exceeded 300 feet and could be seen throughout the industrial cities of Britain. By the late 1800s, when Bancroft wrote, iron smokestacks were often prefabricated and moved to the desired locations. These "cathedrals of industry" were signals that coal was being burned and people were gainfully employed.

In the United States, Pittsburgh, Pennsylvania, capital of steel and iron manufacture, symbolized the progress made possible from burning coal. Again, clear admissions from civic leaders noted the ills of urban pollution created from industries burning coal as early as the late 1800s. Such arguments, however, paled in comparison to the great economic development that fossil fuels made possible. These industrial cities were hubs for dispersed industrial development that could be found tying the nation together.

## RAILROADS LINK ENTERPRISES

In addition to stimulating the development of mining in locales such as Pennsylvania, industrial development contributed to, and fed the development of, related undertakings. More and more industries became essential to everyday American lives. In most cases, each of these undertakings derived from burning coal or other fossil fuels, thereby broadening the impact of emissions. Throughout American history, transportation was one of the most important applications of energy use. In the case of coal, the use of the railroad made coal supplies accessible while also involving coal's energy in innumerable other activities during the 1800s.

The planning and construction of railroads in the United States progressed rapidly during the 19th century. Some historians take the view that this occurred too rapidly. With little direction and supervision from the state governments that were granting charters for construction, railroad companies constructed lines where they were able to take possession of land or on ground that required the least amount of alteration. The first step to any such development was to complete a survey of possible passages.

Before 1840, most surveys were made for the construction of short passenger lines that proved to be financially unsustainable. Under stiff competition from canal companies, many lines were begun only to be abandoned when they were only partially completed. The first real success came in the 1830s, when the Boston and Lowell Railroad succeeded in diverting traffic from the Middlesex Canal. After the first few successful companies demonstrated the economic feasibility of transporting commodities via rail, others followed throughout the northeastern United States.

The process of constructing railroads began with humans reconstruing their view of the landscape. Issues such as grade, elevation, and passages between mountains became part of a new way of mapping the United States. Typically, early railroad surveys and their subsequent construction were financed by private investors. When shorter lines proved successful, investors began talking about grander schemes. These expansive applications of the railroad provided the infrastructure for remarkable commercial growth in the United States, expanding the impact of the Industrial Revolution (Stilgoe, 3–8).

By the 1850s, the most glaring example of this change was coal-powered railroads. The expanding network of rails allowed the nation to expand commercially. Most important, coal-powered railroads knitted together the sprawling United States into a cohesive social and commercial network. Although this could be seen in microscopic examples, including cities such as Pittsburgh and Chicago, to which railroads brought together the raw materials for industrial processes such as steel making, on the macroscopic scale railroads allowed American settlement to extend into the western territories (Stilgoe, 1983).

It was a cruel irony that the industrial era that evolved in the late 1800s relied intrinsically on transportation. Long, slender mountains stretched diagonally across Appalachian regions such as Pennsylvania, creating an extremely inhospitable terrain for transporting raw materials. To access these isolated regions was the goal of a generation of politicians and capitalists. They used their influence to underwrite a network of transportation allowing coal supplies to be accessed and then moved to the sites where energy was needed. This capability made the coal revolution possible. Although the process began with canals, the superior flexibility of railroads soon made them the infrastructure of the industrial era. Rails knit together the sources of raw materials for making iron, steel, and other commodities, allowing the convergence of ingredients in crucial processing centers. The Iron Horse was both the process and product of industrialization (Cronon, 1991b). Anthracite-fueled furnaces were used to make iron rails that were then used to formalize transportation routes throughout the nation. In addition to powering industry, coal became an increasingly popular source for heating in urban residences. There could never be enough coal to fuel such an industrial system, and thus there could never be enough miners and laborers in the mines.

Although each of these social and cultural impacts of the railroad altered American life, the railroads were, after all, primarily economic enterprises. Primitive as it was, the antebellum railroad entirely remade American commerce, particularly reconfiguring the very ideas of prices and costs. Previously, prices had factored in the length of time involved in transporting goods via turnpikes, the steamboat, and the canal. From the start, railroad rates were significantly cheaper than wagon rates. The increasing systemization of the railroad process made low costs even more possible (Cronon, 1991b).

The possibility of railroads connecting the Atlantic and Pacific coasts was soon discussed in the U.S. Congress, and this initiated federal efforts to map and survey the western United States. A series of surveys showed that a railroad could follow any one of a number of different routes. The least expensive appeared to be the 32nd parallel route. The Southern Pacific Railroad was subsequently built along this parallel, even though each constituency tried to argue the worth of routes that benefited themselves. To make this process a bit less politically volatile, the Railroad Act of 1862 put the support of the federal government behind a transcontinental railroad. Creating the Union Pacific Railroad, the act culminated in the linking of the continent when the Union Pacific Railroad's rail met that of the Central Pacific at Promontory, Utah, on May 10, 1869.

Railroading became a dominant force in American life in the late 19th century, and one of its strongest images was its ability to remake the landscape of the entire country. After 1880, the railroad industry reshaped the American built environment and reoriented American thinking away from a horse-drawn past and toward a future with the iron horse.

## STEEL MANUFACTURE PROVIDES INFRASTRUCTURE

Railroads and the reliance on burning fossil fuels enabled the implementation of complex industrial undertakings at a scope and scale never seen before. Although iron manufacture increased in scale with the more intense model of industrialization after 1850, steel is possibly the best example of this new era's capabilities. Using railroads as its linking device, Andrew Carnegie perfected the process of steel manufacturing and created one of the greatest fortunes in history (Opie, 1998).

Into one pound of steel, observed Carnegie, went two pounds of iron ore brought 1,000 miles from Minnesota, 1.3 pounds of coal shipped 50 miles to Pittsburgh, and one-third of a pound of limestone brought 150 miles from Pittsburgh. Rivers and railroads brought the material to the Carnegie Steel Works along Pittsburgh's Monogahela River where Bessemer blast furnaces fused the materials into steel. One of the greatest reasons for the rapid rise of American industry was its flexibility compared to that of other nations. Railroading could be integrated immediately into various industries in the United States, which, for instance, allowed American industry to immediately embrace the new Bessemer steel-making-technology. Other nations, such as Britain, needed to shift from previous methods.

One innovation contributed to another in the late industrial era. Inexpensive energy made it feasible to gather the disparate materials that were necessary to make steel. Steel was stronger and more malleable than iron, which made possible new forms of building. Carbon levels account for the bulk of the distinction between the two metals. Experiments with removing the oxygen content of pig iron required more heat than ordinary furnaces could muster. The Bessemer invention created a "Bessemer blow," which included a violent explosion to separate off additional carbon and produce the 0.4 percent oxygen level that was desirable for steel.

New tasks, such as running the Bessemer furnace, created specialized but also very dangerous jobs. Working in the steel mill created a new hierarchy for factory towns. In the case of steel-making, hot or dangerous jobs such working around the Bessemer furnace eventually fell to African American workers (Opie, 1998).

## ELECTRICITY AND THE EVOLUTION OF THE ENERGY INDUSTRY

Industrial applications of energy shaped the industrial era; however, by later in the 1800s, coal, in the form of electricity, was also remaking the everyday lives of many Americans. On the whole, new energy made from fossil fuels altered almost every American's life by 1900. In 1860, there were fewer than a million and a half factory workers in the country; by 1920, there were 8.5 million. In 1860, there were about 31,000 miles of railroad in the United States; by 1915, there were nearly 250,000 miles. The energy moving through such infrastructure would not remain limited to the workplace.

In the 19th century, energy defined industry and work in America, but it did not necessarily impact everyday cultural life. This would change dramatically by the end of the 1800s with the development of technology to create, distribute, and put to use electricity. Although electricity is the basis for a major U.S. energy industry, it is not itself an energy source. It is mostly generated from fossil fuel (coal, oil, natural gas), hydroelectric (water power), and nuclear power. The electric utilities industry includes a large and complex distribution system and, as such, is divided into transmission and distribution.

After experiments in Europe, the United States's electrical future fell to Thomas Edison, one of the nation's great inventors. The 1878 discovery of the incandescent filament lamp actually is credited to British scientist Joseph Swan; however, within 12 months Edison made a similar discovery. Starting with his laboratory and then, in September 1882, a New York City street, Edison used his DC generator to provide artificial lighting. Subsequently, George Westinghouse patented a motor for generating alternating current. Society became convinced that its future lay with AC generation. This, of course, required a level of infrastructural development that would enable the utility industry to have a dominant role in American life.

Once again, this need for infrastructural development also created a great business opportunity. Samuel Insull went straight to the source of electric technology and ascertained the business connections that would be necessary for its development. He got to know the business from the inside, when, in 1870, he became a secretary for George A. Gourand, one of Thomas Edison's agents in England. In 1881, Insull moved to the United States to become Edison's personal secretary (Hughes, 226–30). Continuing with Edison, Insull was vice president of Edison General Electric Company in Schenectady, New York by 1889. When financier J. P. Morgan took over Edison's power companies in 1892, he sent Insull to Chicago to assist the struggling Chicago Edison Company. It was in Chicago that Insull created the next great revolution in the industry. After the economic panic of 1893, Insull oversaw Chicago Edison's expansion as it bought out its competitors cheaply. With this control, he centralized the power grid by constructing a large central power plant along the Chicago River at Harrison Street.

By 1908, Insull's Commonwealth Edison Company made and distributed all of Chicago's power. Insull connected electricity with the concept of energy and also diversified into supplying gas. Then he pioneered the construction of systems of dispersing these energy sources into the countryside. The energy grid was born. It would prove to be the infrastructure behind each American life in the 20th century. Through the application of this new technology, humans now could defy the limits of the sun and season (Hughes, 234–40).

The greatest application of electrical technology—and a symbol of humans' increased reliance on fossil-fueled power—is the light bulb. For decades, inventors and businessmen had been trying to invent a source of light that would be powered by electricity. Primarily, their experiments emphasized positioning a filament in a vacuum. An electric current then was sent through in hopes of making the filament glow. The filaments consistently failed, however, disintegrating as soon as the current reached them (Hughes, 39–40).

In 1878, Thomas Edison decided to concentrate his inventive resources on perfecting the light bulb. Instead of making his filament from carbon, Edison switched to platinum, which was a more resilient material. In addition, in 1879, he began using the Sprengel vacuum, which was an improved vacuum pump. With this more effective pump, Edison was able to modify his design and implement the use of the less-expensive carbon filaments. Using a carbonized piece of sewing thread as a filament, in late October 1879, Edison succeeded in keeping a lamp burning for 13.5 hours. He later modified the filament to be horseshoe-shaped, which eventually supported burning times in excess of 100 hours. Edison had invented a practical light bulb; but, more important, he cleared the path for the establishment of the electrical power system that would revolutionize human existence.

It was this power system that became Edison's real achievement and created the market that would beget a huge new industry destined to affect the lives of every American. The nature of everyday life became defined by activities made possible by electric lighting as well as the nearly endless amount of other electrically powered items. The light bulb was a critical innovation in the electrification of America; however, it also helped to create the market that stimulated efforts to perfect the industry of power generation (Nye, 1999, 138–42).

At the root of power generation, of course, was the dynamo. The dynamo was the device that turned mechanical energy of any type into electrical power (Nye, 1999, 144–8). Even while Edison worked to design a better light bulb, the dynamos generating the electricity had only reached approximately 40 percent of their possible efficiency. In his lab, Edison created a dynamo that raised this to 82 percent.

Together, Edison's innovations established electricity's flexibility, which he demonstrated in September 1882 from the central station on Pearl Street in Manhattan. Shortly thereafter, his plant supplied electricity to a one-mile square section of New York City, creating a model for the nation. Such areas became futuristic symbols for the growing nation. In applying electricity to domestic, everyday uses, Edison and others had done a remarkable job of separating benign applications, such as powering a refrigerator, from the coal-burning necessary to release the required energy. This abstraction became ever more pronounced as Americans used such power in more diverse ways in the

20th century, adding many layers between our applications of energy and the emissions required to generate it.

## A NEW SCALE AND SCOPE FOR ENERGY: BLACK GOLD

Coal provided the basic underpinning for the Industrial Revolution. Through its impact on the factory system, it radically changed American life. But a similar dependence derived from another primary energy resource that followed in the wake of coal: petroleum. The degree to which American life in the 21st century was bound up with petroleum would have shocked 19th-century users of "Pennsylvania rock oil." Most farmers who knew about the oil in the early 1800s viewed seeping crude only as a nuisance to agriculture and a polluter of water supplies. These observers were not the first people to consider the usefulness of petroleum (or lack thereof); petroleum had been a part of human society for thousands of years. Yet its value grew only when European-Americans applied their commodity-making skills to the resource.

As oil's reputation rose, settlers in northwestern Pennsylvania gathered oil from springs on their property by constructing dams of loose stones to confine the floating oil for collection. In the mid-1840s, one entrepreneur, Samuel Kier, noticed the similarity between the oil prescribed to his ill wife and the annoying substance that was invading the salt wells on his family's property outside Pittsburgh. He began bottling the waste substance in 1849 and marketed it as a mysterious cure-all throughout the northeastern United States. Although he still acquired the oil only by skimming, Kier's supply quickly exceeded demand because there was a constant flow of the oil from the salt wells. With the excess, he began the first experiments with using the substance as an illuminant, or substance that gives off light. The culture of expansion and development was beginning to focus on petroleum (Black, 2000).

From this point forward, petroleum became the product of entrepreneurs—except for one important character: Edwin L. Drake of the New Haven Railroad. In 1857, the company sent Drake to Titusville, Pennsylvania, to attempt to drill the first well intended to extract oil. The novelty of the project soon wore off for Drake and his assistant, Billy Smith. The towns-people irreverently heckled the endeavors of a "lunatic." During the late summer of 1859, Drake ran out of funds and wired to New Haven, Con-necticut, for more money. He was told that he would be given money only for a trip home—that the Seneca Oil Company, as the group was now called, was done supporting him in this folly. Drake took out a personal line of credit to continue, and a few days later, on August 29, 1859, Drake and his assistant discovered oozing oil.

Throughout its history, petroleum has exhibited wide fluctuations in price and output. The boom and bust cycle was even underwritten by the courts in the case of *Brown v. Vandergrift* (1875), which established the laissez-faire development policy that became known as "the rule of capture." The oil could be owned by whomever first pulled it from the ground, that is, captured it. The rush to newly opened areas became a race to be the first to sink the wells that would bring the most oil up from its geological pockets (Black, 2000). After the American Civil War, the industry consistently moved toward the streamlined state that would allow it to grow into the world's major source of energy and lubrication during the 20th century.

During the 19th century, petroleum's most significant impact may have been on business culture. The culture of the industry that took shape would change land use and ideas of energy management throughout the world. John D. Rockefeller and Standard Oil first demonstrated the possible domination available to those who controlled the flow of crude oil. Rockefeller's system of refineries grew so great at the close of the 19th century that he could demand lower rates and eventually even kickbacks from rail companies. One by one, he put his competitors out of business and his own corporation grew into what observers in the late 1800s called a trust (what, today, is called a monopoly). Standard's reach extended throughout the world, and it became a symbol of the Gilded Age when businesses were allowed to grow too large and benefit only a few wealthy people. Reformers vowed things would change (Chernow, 1998).

The laissez-faire era of government regulation of businesses, particularly energy companies such as Standard, came to an end when Progressive reformers pressed for a different view of the government's role in American life. President Theodore Roosevelt, who took office in 1901, led the Progressive effort to involve the federal government in monitoring the business sector. In the late 1890s, "muckraking" journalists had written articles and books that exposed unfair and hazardous business practices. Ida Tarbell, an editor at *McClure's* magazine who had grown up the daughter of a barrel maker in Titusville, took aim at Rockefeller. Her *History of the Standard Oil Company* produced a national furor over unfair trading practices. Roosevelt used her information to enforce antitrust laws, resulting in Standard's dissolution in 1911. Rockefeller's company had become so large that when broken into subsidiaries, the pieces would grow to be Mobil, Exxon, Chevron, Amoco, Conoco, and Atlantic, among others (Tarbell, 2003).

Even after Standard's dissolution, the image of its dominance continued. Standard had led the way into international oil exploration, suggesting that national borders need not limit the oil-controlling entity. Throughout the 20th century, large multinational corporations or individual wealthy businessmen attempted to develop supplies and bring them to market. Their efforts meshed with consumer desire to make petroleum the defining energy resource of the 20th century. As with coal, however, the real revolution in consumption required basic changes in supply and in the scale and scope of petroleum use in American life.

## CHEAP OIL SETS THE TONE FOR OUR HIGH-ENERGY EXISTENCE

The revolution in the supply of petroleum began with international expansion; however, it was a domestic source that truly defined petroleum's role in Americans' high-energy existence. Although new drilling technologies helped to increase supply, entire new regions were required to sustain the industry's development. By 1900, companies such as Standard Oil sought to develop new fields all over the world. In terms of the domestic supply of crude oil, however, the most significant breakthrough came in Texas. With one 1901 strike, the limited supply of crude oil became a thing of America's past. It is no coincidence, then, that the century that followed was powered by petroleum.

This important moment came in East Texas, where, without warning, the level plains near Beaumont abruptly give way to a lone, rounded hill before returning to flatness. Geologists call these abrupt rises in the land "domes" because hollow caverns lie beneath. Over time, layers of rock rise to a common apex and create a spacious reservoir underneath. Salt often forms in these empty geological bubbles, creating a salt dome. Over millions of years, water or other material might fill the reservoir. At least, that was Patillo Higgins's idea in eastern Texas during the 1890s. Higgins and very few others imagined such caverns as natural treasure-houses. Higgins grew intrigued with one dome-shaped hill in southeast Texas. Known as Spindletop, this salt dome, with Higgins's help, would change human existence.

Texas had not yet been identified as an oil producer. Well-known oil country lay in the eastern United States, particularly in western Pennsylvania. Titusville, Pennsylvania had introduced Americans to massive amounts of crude oil for the first time in 1859. By the 1890s, petroleum-derived kerosene had become the world's most popular fuel for lighting. Thomas Edison's experiments with electric lighting placed petroleum's future in doubt; however, petroleum still stimulated a boom wherever it was found. But in Texas? Every geologist who inspected the "Big Hill" at Spindletop told Higgins that he was a fool.

With growing frustration, Higgins placed a magazine advertisement requesting someone to drill on the Big Hill. The only response came from Captain Anthony F. Lucas, who had prospected domes in Texas for salt and sulfur. On January 10, 1901, Lucas's drilling crew—known as "roughnecks" for the hard physical labor of drilling pipe deep into Earth—found mud bubbling in their drill hole. The sound of a cannon turned to a roar, and suddenly oil spurted out of the hole. The Lucas geyser, found at a depth of 1,139 feet, blew a stream of oil over 100 feet high until it was capped nine days later. During this period, an estimated 100,000 barrels of oil flowed from the well per day—well beyond any flows previously witnessed. By the time Lucas gained control of the geyser on January 19, the massive pool of oil had attracted not only gawkers but competitors, quickly transforming Beaumont into Texas's first oil boomtown.

The flow from this well, named Lucas 1, was unlike anything witnessed before in the petroleum industry: 75,000 barrels per day. As news of the gusher reached around the world, the Texas oil boom was on. Land sold for wildly erratic prices. After a few months more than 200 wells had been sunk on the Big Hill. By the end of 1901, an estimated $235 million had been invested in oil in Texas. This was the new frontier of oil; however, the industry's scale had changed completely at Spindletop. Unimaginable amounts of petroleum—and the raw energy that it contained—were now available at a low enough price to become part of every American's life.

It was the businessmen who then took over from Higgins and other petroleum wildcatters. Rockefeller's Standard Oil and other oil executives had managed to export petroleum technology and exploited supplies worldwide. The modern-day oil company became a version of the joint-stock companies that had been created by European royalty to explore the world during the period of mercantilism of the 1600s. Now, behemoth oil companies were transnational corporations, largely unregulated and seeking one thing: crude oil. Wherever "black gold" was found, oil tycoons set the wheels of development in motion. Boomtowns modeled after those in the Pennsylvania oil fields could suddenly pop up in Azerbaijan, Borneo, or Sumatra (Yergin, 117–19).

As East Texas gushers created uncontrollable lakes of crude, no one considered the idea of shortage or conservation. Even the idea of importing oil was a foreign concept. California and Texas flooded the market with more than enough crude oil, and then, from nearly nowhere, Oklahoma emerged in 1905 to become the nation's greatest oil producer. Now, however, the question became what was to be done with this abundant, inexpensive source of energy.

## MAKING BLACK GOLD FROM TAR

The second key to petroleum's expansion was for it to become essential. This basic utility for petroleum came through transportation. Commodities such as petroleum are culturally constructed: a market must first place a value on them before they are perceived as worthwhile. As had not been the case with coal, Americans ultimately formed a very personal, cultural relationship with petroleum during the 20th century. In the earliest years of petroleum, it was refined into kerosene, an illuminant to replace whale oil. After 1900, when electricity became the source of most lighting, petroleum's greatest value derived from its use in transportation, mainly the automobile. The amounts of crude available after 1900 made it cost-effective to use petroleum in new ways: particularly, to burn it in engines and motors. But just as with burning coal, burning petroleum as a fuel releases gases into the atmosphere.

First developed in Europe in the late 1800s, the automobile was marketed successfully beginning in 1894. Inconvenience from a lack of roads and infrastructure precluded Americans from rapidly accepting the new "horseless carriage." Although independent mechanics and inventors created many early vehicles, it was Ransom E. Olds who brought mass production to the automobile industry, in 1901. Selling each vehicle for $650, Olds's company turned out more than 400 Oldsmobiles in its first year. Of course, his innovation cleared the way for competitors Henry M. Leland and Henry Ford to expand mass production methods applied to the automobile, but who actually owned this evolving technology remained unclear. Through court battles that stretched into the 1910s, aspiring automobile manufacturers debated who owned the patents for the basic technologies of the automobile, ranging from the internal combustion engine to the mass production process itself. Partly as a result of this litigation, the Ford Motor Company organized in 1903, the General Motors Corporation in 1908, and the Chrysler Corporation in 1925 (McShane, 1994).

The manufacturing and marketing efforts of Henry Ford and others changed the American attitude toward the automobile. By 1913, there was one motor vehicle to every eight Americans. Ford's Model T, nicknamed the "flivver" and the "tin lizzie," proved to be the vehicle that changed the industry. Ford had mass-produced the first one in 1908, and over the next 20 years, his company sold 15 million Model Ts. World War I had a great deal to do with changing the market for automobiles. Although production was virtually halted during the war, afterward the motorcar emerged as more than a novelty in the expanding United States. Most of the basic mechanical difficulties of the auto had been solved, and it truly had become an incredible convenience. With the basic kinks gone, manufacturers focused on making vehicles safer and, by improving style and comfort, more desirable. Even Ford responded to the competitive pressure, adjusting his Model T and ultimately releasing the Model A in October 1927 (Brinkley, 2003).

Ford's model of mass production ensured that by the 1920s, the car had become no longer a luxury, but a necessity, of American middle-class life. The need for additional infrastructure (roads and bridges) was growing, but it was unclear who would pay to develop it (Lewis, 1997). The Federal Road Act of 1916 cleared the way for taxpayer dollars to be put to use to construct roads and set the stage for one of the most significant landscape transformations in human history. Created in the 1920s, the Bureau of Public Roads was empowered to plan a highway network to connect all cities of 50,000 or more inhabitants. Some states acquired additional funding through taxes on the gasoline that drivers needed to put in their tanks. Finally, during the 1950s, President Dwight D. Eisenhower dedicated federal funding to a national system of large highways that would, it was asserted, assist urban centers in preparedness for nuclear attack. Ultimately, the Interstate Highway Act created a national system of roads that made the United States the world's model petroleum-based society (Jackson, 1985).

In the United States, roads initiated related social trends that added to Americans' dependence on petroleum. Most important, between 1945 and 1954, 9 million people moved to suburbs. The majority of the suburbs were connected to urban hubs only by the automobile. Between 1950 and 1976, central city population grew by 10 million, while suburban growth was 85 million. Housing developments and the shopping/strip mall culture that accompanied decentralization of the population made the automobile a virtual necessity. Shopping malls, suburbs, and fast-food restaurants became the American norm through the end of the 20th century, making American reliance on petroleum complete. Americans now were entirely wedded to their automobiles, which allowed prices of petroleum to impact life in the United States more than in any other nation (Lewis, 1997).

Although the gravity with which emissions altered Earth's air was not fully appreciated until much later, pollution from the internal combustion engine at the ground level was criticized from the start. To appreciate how these localized issues, such as smog, might be indicators of eventual problems of a much larger scale required scientists to achieve a series of accomplishments during the 20th century. Primary among these was perspective: the ability to rise above the ground level on which humans live and to perceive the movement of air on a global scale.

## A NEW VIEW OF EARTH AND THE HIGH-ENERGY LIFE

If the human view of the Earth had remained unchanged, it is very possible that the science of global warming would have remained incomprehensible.

Therefore, both in terms of public reaction and in terms of the science to establish climate change, the view of Earth from space was critical.

As humans traveled to space for the first time in history during the 1960s, their view of the universe changed, but so did their view of their role in it. During each of the early trips, astronauts photographed Earth from space. In snapping these photos, they captured a perspective no human had ever before seen: Earth as an organic object, devoid of political divisions and any evidence of human habitation. The first photo, made public in 1968, was referred to as "Earthrise." "Whole Earth" followed on December 7, 1972, during the Apollo 17 mission. Together, these images profoundly altered human ideas of Earth. "Whole Earth" reached an American public ready to reconsider its place in the world and led to the environmental campaign: "Think Globally, Act Locally."

"Whole Earth" put a biological face on the planet. The view from space revealed identifiable environments without differentiating by border, politics, or economics. In this "Whole Earth" view, the globe, more than ever before in human history, emerged as much greater than the humans who resided on it. Some geographers specifically noted that by centralizing the view of locations such as Africa, the Middle East, and Antarctica, the Apollo 17 photograph "upsets conventional Western cartographic conventions." Instead of a U.S.-centered view for Americans, "Whole Earth" revealed a global natural environment.

"Whole Earth" offered a view of Earth that played directly into the growing culture of environmentalism. The organic Earth portrayed in this photo appeared as an environmentally threatened home. Activists such as Stewart Brand embraced this profoundly new perspective of Earth. Space travel, in fact, offered a type of technological middle ground: a perspective brought by new technology that afforded us a better understanding of the world around us. Clearly, this technology was a tool for global good, unlike the more ambiguous nuclear technology with its potentially catastrophic possibilities.

In fact, "Whole Earth" marked a new approach for NASA's use of technology. In addition to the initiatives that would make space the next frontier for American society, NASA increasingly used its research to examine and understand nature back on Earth. Soon after Apollo 17 astronauts took the "Whole Earth" picture, NASA launched its first Landsat satellite to help examine ecological changes from soil erosion and deforestation to ocean dumping, air pollution, and urban sprawl (Maher, 2004). In environmental thought, such a viewpoint helped to inspire scientist James Lovelock to propose the "Gaia hypothesis," arguing that Earth was a living organism that demanded proper treatment. Many observers of an image such as "Whole Earth" might admit that although Gaia does not have to be taken literally, it presents the basic idea of Earth as a complex system of intertwined feedbacks.

This perspective helped motivate some humans to eventually change their attitude toward such actions on Earth as the burning of fossil fuels, which had become entirely acceptable over the course of the industrial revolution. Ironically, of course, it was technological development, such as rocket engines and space travel, that brought us this new perspective.

The high-energy lifestyle that humans adopted during the 20th century has enabled staggering accomplishments and advances, including the view from space. Extensive use of fossil fuels has also come with its share of serious impacts, some of which we are just becoming aware of in the 21st century. This chapter began with humans at the time of the Agriculture Revolution, living within the natural cycles of energy that begin with the sun. The Industrial Revolution made such natural forms of energy merely alternatives to the primary use of energy made from burning fossil fuels. Particularly in the United States, we made cheap energy part of our life and threw caution to the wind. The 20th century became a binge on cheap energy and all that it made possible.

At the dawn of the 21st century, Americans have come to recognize that the great energy resources of the industrial era were exhaustible, that the supplies of coal, petroleum, and natural gas were finite, and that, in addition, the emissions from burning and utilizing these resources created impacts on a global scale. Throughout the 20th century, amidst the frenzy of energy consumption and its associated economic and social development, a growing chorus had attempted to alert consumers and politicians to the temporality of reliance on hydrocarbon-derived energy. For these voices, the growing scientific appreciation of the impacts of industrialization provided a bold new rationale for pursuing alternative lines of development.

# 4

# Domestic Politics and Ethics Regarding the Environment

The "American Century" is how Henry Luce, founder of *Life* magazine, described the 20th century once Americans had begun to form the post–World War II world. This moment of exaggeration seems understandable in the post-1945 world that saw the United States serve as the decisive force among the Allies and also emerge as the least scathed by war of any of the major participants. Finally, the view of the century was also influenced by a remarkably brief—approximately four-year—period in which the United States alone could boast the possession of functioning atomic weapons.

The true power behind the maturation of the "American Century" concept, however, was the standard of living made possible and even compelled by the military and security situation of the latter half of the century. And in this worldview, a guiding model for the developed world was the standard of living that became known as "conspicuous consumption." The lifeblood of our consumer nation was cheap energy, primarily gotten by burning fossil fuels. In the spirit of the Cold War, such development was a patriotic symbol of a new America that would stand as a model for the developing world.

To suggest problems or shortcomings of this model of continuous growth and expansive development was construed by some Americans as unpatriotic and even immoral. Steeped in some of this same cultural baggage, the

This chapter is based in part on Brian C. Black's "The Consumer's Hand Made Visible: Consumer Culture in American Petroleum Consumption of the 1970s," an unpublished paper delivered at the Conference on Energy in Historical Perspective: American Energy Policy in the 1970s, conference, University of Houston, November 9–10, 2007.

discourse over climate change often drifts toward questions of morality: one side claiming that it is immoral to continue practices that scientific evidence deemed damaging to the environment; another side claiming that it is immoral for the United States to compromise its competitive advantage in economic development by demanding shifts to our most basic infrastructure. All critical issues in American society are not so volatile; climate change seems to access nerves that are already inflamed by existing cultural and political disagreements.

In recent years, this debate veered toward the apocalyptic and, therefore, often incorporated religion and basic ideas of God and humans' place among Earth's various inhabitants. In particular, the climate change debate has led to significant changes in modern environmentalism, including two extremely different developments: first, the broadening of scientific understanding and interest in environmentalism, and second, the demonizing of environmental perspectives as antidevelopment and unpatriotic. For all of these reasons, passions run higher in this debate than nearly any other.

In this chapter, we strive to offer each perspective objectively. Whenever possible, we will offer extracts or excerpts from speeches or writings to represent each perspective. The first topic may be the most complex of all of these and, thus, it will require the most explanation. This point, in short, begins with the realization that the ability of proponents of the hypothesis of global warming to seize the ears of policymakers and, eventually, a large segment of the human population did not occur by chance. If placed in the larger context of changing ideas of the human relationship with nature, the idea of global warming can be seen as being partly enabled by the evolution of modern environmentalism, a political and social movement evolving in many nations after the 1960s.

## SHADES OF GREEN

The intellectual roots of American environmentalism are most often traced back to 19th-century ideas of romanticism and transcendentalism and thinkers such as Henry David Thoreau. These ideas of conservation or preservation are important components of what scholars today refer to as modern environmentalism; however, today's environmentalism composes a social movement with clear political outcomes. Most scholars include modern environmentalism with other reform efforts growing out of the 1960s.

Overall, the 1960s counterculture contributed to the development of many institutions that would change basic relationships in American life. The American relationship with nature was one of the most prominent shifts. Much of what became known as the modern environmental movement was organized around groups and organizations that prospered with the influence of 1960s radicalism; however, the real impact of these organizations came during the later 1960s and 1970s when their membership skyrocketed with large numbers of the concerned, not-so-radical, middle class. These growing organizations demanded a political response from lawmakers. Whether the issue was pollution, the need for alternative energy, or the overcutting of rain forest, concerned citizens exerted an environmental perspective on a scale not seen previously.

Contrasted with the conservation movement of the late 19th century, the social landscape of 20th century environmentalism had changed a great deal. For instance, many of these environmental special-interest groups would evolve into major political players through lobbying. Nongovernmental organizations (NGOs) broadened the grassroots influence of environmental thought; however, they also created a niche for more radical environmentalists. The broad appeal as well as the number of special-interest portions of environmental thought stood in stark contrast to 19th-century environmentalism. Whereas early conservationists were almost entirely members of the upper economic classes of American society, the new environmentalists came mostly from the middle class that grew rapidly after World War II (Opie, 418–25).

During the 1970s and 1980s, these NGOs helped to bring environmental concern into mainstream American culture. Some critics argue that American living patterns changed little; however, the awareness and concern over human society's impact on nature had reached an all-time high in American history. These organizations often initiated the call for specific policies and then lobbied members of Congress to create legislation. By the 1980s, NGOs had created a new political battlefield as each side of environmental arguments lobbied lawmakers. Environmentalists were often the ones offering their perspective as the moral high-road that would help rein in human development and expansion with an ethic of restraint.

The American public often financially supported organizations that argued for their particular perspectives. Even traditional conservationist organizations such as the Sierra Club (1892), National Audubon Society (1905), National Parks and Conservation Society (1919), Wilderness Society (1935), National Wildlife Federation (1936), and Nature Conservancy (1951) took much more active roles in policymaking. The interest of such organizations in appealing to mainstream, middle-class Americans helped to broaden the base of environmental activists. It also contributed to the formation of more radical-thinking environmental NGOs that disliked the mainstream interests of the larger organizations. In fact, many devout environmentalists argued that some of these NGOs were part of the "establishment" that they wished to fight.

One of the first writers to take advantage of this increased interest among middle-class Americans was the scientist and nature writer Rachel Carson. She began writing about nature for general readers in the late 1950s. In 1962, Carson's *Silent Spring* erupted onto the public scene to become a bestseller after first being serialized in *The New Yorker*. Carson's scientific findings brought into question basic assumptions that Americans had about their own safety and many of the chemicals that they used to create their comfortable standard of living.

Overall, the cultural attitude toward progress predicated on cheap energy and manufactured chemicals was beginning to lose its dominant hold. In the case of Carson's work, her expose of the health impacts of chemicals helped to disrupt the paradigm that supported Americans' trust, more generally, in technological progress. In a single summer, chemical science and blind confidence in technological progress had fallen from its unchallenged pedestal. Here is a portion of what Carson wrote:

The "control of nature" is a phrase conceived in arrogance, born of the Neanderthal age of biology and philosophy, when it was supposed that nature exists for the convenience of man. The concepts and practices of applied entomology for the most part date from that Stone Age of science. It is our alarming misfortune that so primitive a science has armed itself with the most modern and terrible weapons, and that in turning them against the insects it has also turned them against the Earth. (Carson, 12–14)

Her story and her words would inspire an entire portion of the American population to reconsider our society's living patterns.

Following Rachel Carson, in 1968 Garrett Hardin wrote an article that developed the ecological idea of the commons. This concept and his argument of its tragic (undeniable) outcome in depletion gave humans new rationale with which to view common resources such as the air and the ocean. He wrote:

The tragedy of the commons develops in this way. Picture a pasture open to all. It is to be expected that each herdsman will try to keep as many cattle as possible on the commons. Such an arrangement may work reasonably satisfactorily for centuries because tribal wars, poaching, and disease keep the numbers of both man and beast well below the carrying capacity of the land. Finally, however, comes the day of reckoning, that is, the day when the long-desired goal of social stability becomes a reality. At this point, the inherent logic of the commons remorselessly generates tragedy.

As a rational being, each herdsman seeks to maximize his gain. Explicitly or implicitly, more or less consciously, he asks, "What is the utility to me of adding one more animal to my herd?" This utility has one negative and one positive component. . . .

Adding together the components . . . the rational herdsman concludes that the only sensible course for him to pursue is to add another animal to his herd. And another. . . . But this is the conclusion reached by each and every rational herdsman sharing a commons. Therein is the tragedy. Each man is locked into a system that compels him to increase his herd without limit—in a world that is limited. Ruin is the destination toward which all men rush, each pursuing his own best interest in a society that believes in the freedom of the commons. Freedom in a commons brings ruin to all. (Hardin, 243–48)

This essay marked a crucial moment in Americans' ability to apply the scientific ideas of ecology, conservation, and biology to human life—to assume that we possibly were *not* the exceptional species on Earth. Or, even if we were exceptional, maybe this status came with a responsibility for stewardship and management instead of for expansion. Such new perspectives were critical to preparing the cultural soil that would be able to conceive of the concept of global warming when it was presented by scientists. Most important, for the first time Americans were learning to view the world around them as not indestructible but, instead, quite volatile and composed of limited resources.

## ENVIRONMENTAL MOVEMENT ZEROES IN ON THE ICE

By the second half of the 1900s, many activists had become very outspoken that Americans had a villain living among their everyday lives: the internal combustion engine (ICE). In its earliest version, this reimaging of the ICE had little to do with the growing scarcity of petroleum supplies; instead, critics emphasized the inevitable outcome of burning petroleum in car engines: emissions and air pollution. Before the public understood concepts such as greenhouse gasses and climate change, air pollution was acknowledged to be unpleasant and likely unhealthy.

Air pollution had been documented to possess health impacts for humans since the early 1900s. In the 1940s, air pollution was more broadly construed to be a product of many facets of urban life, not just factories. In cities such as Los Angeles, the phenomenon became known as "smog" and was directly connected to exhaust from the automobile's ICE (typically smog is not merely used to blend the words smoke and fog but also to include chemical compounds that linger in the air when sunlight interacts with smoke put off by burning hydrocarbons). The existence of carbon monoxide, carbon dioxide, and sulfur dioxide in such air pollution was not clarified for a few more decades. The connection between smog and auto exhaust is credited to Arie Haagen-Smit, a researcher at the California Institute of Technology. During the 1950s, Haagen-Smit fought off the savage criticism of the auto manufacturers who claimed that a well-tuned vehicle had no such adverse effects on the air. Severe smog episodes in California kept the issue in the public arena and helped to make it one of the primary issues for the nascent environmental movement.

Many activities that related to air pollution were planned for Earth Day 1970, a seminal event in the emergence of modern environmentalism. In one of the day's most dramatic and public displays, New York City's Fifth Avenue was transformed into an auto-free zone. Only pedestrian traffic was allowed to traverse the city's symbolic primary artery, revealing by contrast the noise, congestion, and exhaust that the vehicles normally brought to the space.

As scientists began to understand the complexities of air pollution in the late 1960s, it became increasingly apparent that in addition to specific toxic emissions such as lead, the ICE was a primary contributor to air pollution and smog. It has been estimated that emissions from the nation's nearly 200 million cars and trucks account for about half of all air pollution in the United States, and in cities this estimate jumps to more than 80 percent. The American Lung Association estimates that America spends more than $60 billion each year on health care as a direct result of air pollution (Doyle, 134).

When the engines of automobiles and other vehicles burn gasoline, they create pollution. These emissions have a significant impact on the air, particularly in congested urban areas. This impact is hard to track or trace, however, because the sources are moving. The pollutants included in these emissions are carbon monoxide, hydrocarbons, nitrogen oxides, and particulate matter. Nationwide, mobile sources represent the largest contributor to air toxins, which are pollutants known or suspected to cause cancer or other serious health effects. These are not the only problems. Internal combustion engines also emit greenhouse gases, which scientists believe are responsible for trapping heat in the Earth's atmosphere.

Initial efforts at controlling auto emissions began at the state level. These efforts date back to 1961, when California exceeded anything being considered on the national level and required all cars to be fitted with PCV valves that helped contain some of the emissions within the vehicle's crankcase. Federal legislation began in 1965 with the Motor Vehicle and Air Pollution Act, followed in 1970 by the Clean Air Act. As the new social movement of modern environmentalism took shape after Earth Day 1970, however, constituents forced many lawmakers to consider drastic changes to our vehicles.

The organizer of Earth Day, Gaylord Nelson, in fact, went on record in 1970 saying that: "The automobile pollution problem must be met head on with the requirement that the internal combustion engine be replaced by January, 1, 1975" (Doyle, 64). The 1973 Oil Embargo added supply concerns to the calls for the construction of more efficient engines. Together, an aggregate of concerns focused public opinion against the ICE as both inefficient and a polluting threat to U.S. health and security. Although Gaylord Nelson and others argued for banning the engine altogether, the most likely outcome appeared to be new federal controls on emissions (similar to those used in California) that would apply to all American cars. These new standards or controls were called Corporate Average Fuel Economy CAFÉ standards, and their acceptance was not seen as a moral good by everyone.

The battle over how far CAFÉ standards and emissions controls would extend motivated the American auto industry to flex its political muscle to a degree it never had before. Very quickly, the health and safety concerns being voiced morphed into threats of inflated prices on American cars and the dire economic prospect of foreign autos encroaching on the American market. After meeting with President Richard Nixon during 1972–73, industry leaders altered their approach. When they met with President Gerald Ford in 1975, the auto industry executives offered to accept a 40 percent improvement in mileage standards if, in return, Congress would ease standards on emissions. Ford agreed and presented this compromise to American consumers in his State of the Union address. Although Congress protested, this approach of accepting CAFÉ standards while relaxing emissions standards became the rallying point for the auto industry during the 1970s.

The complex terrain of the policies relating to automobiles demonstrate the technology's primacy in America's social and economic life. Auto emissions were one of the first emphasis of environmental policy, with initial legislation passed in 1970 (National emission standards were contained in the Clean Air Act). One of the major proponents of clean air legislation was Senator Edwin Muskie, a Democrat from Maine, who bridged the concerns of the new environmental NGOs that had sprung up from middle-class America's Earth Day exuberance and the 1960s notion that an activist federal government could regulate and ultimately solve the nation's various ills. As the details of the Clean Air Act were worked out in Congress, Muskie won a major victory when specific pollutants contained in vehicle exhaust, such as CO and HCl, were required to decrease 90 percent from 1970 levels by 1975. The intention, of course, was to force manufacturers to create the technologies that could meet the new standards. Individual states led the way. In 1975, a California act required that vehicle exhaust systems be modified prior to the muffler to include a device called the catalytic converter. Costing approximately $300, early converters ran the exhaust through a canister of pellets or honeycomb made of either stainless steel or ceramic. The converters offered a profound, cost-effective way of refashioning the existing fleet of vehicles to accommodate new expectations on auto emissions.

In addition, the scientific scrutiny of auto emissions proceeded on one additional, much more specific front. Air testing on emissions and the smog that they created also revealed a now undeniable reality of auto use: lead poisoning. The willingness to tolerate lead additives in gasoline had persisted from the 1920s. Under the new expectations of the 1970s, lead emissions presented auto manufacturers with a dramatic change in the public's expectations.

By this point, the amount of lead added to a gallon of gasoline "hovered in the vicinity of 2.4 grams. The Department of Health, Education and Welfare, which was home to the Surgeon General starting with the Kennedy Administration, had authority over lead emissions under the Clean Air Act of 1963. The criteria mandated by this statute were still in the draft stage when the Act was reauthorized in 1970 and a new agency called the Environmental Protection Agency (EPA) came into existence" (Lewis [1985], cited in Faulk and Gray 2006, 93). As just one indicator of changing expectations and ethics of morals among the American public, the days of lead's use in American gas tanks was clearly on the wane.

In January 1971, EPA's first administrator, William D. Ruckelshaus, declared that "an extensive body of information exists which indicates that the addition of alkyl lead to gasoline . . . results in lead particles that pose a threat to public health." The resulting EPA study, which was released on November 28, 1973, confirmed that lead from automobile exhaust posed a direct threat to public health. The EPA subsequently issued regulations calling for a phased reduction in the lead content of the nation's total gasoline supply, applying to all grades of gasoline. Following California's lead with catalytic converters, U.S. automakers responded to EPA's lead-reduction timetable by equipping cars manufactured in 1975 and subsequent years with pollution-reducing catalytic converters that required the use of unleaded fuel. With the fleet largely converted by 1989, Congress finally banned the use of leaded gasoline entirely (Gorman, 2001). It is estimated that from the 1920s, when manufacturers had convinced Americans that lead (called Ethyl) was a safe additive to gasoline, until 1989, 15.4 billion pounds of lead dust were spewed into the air by automobiles (Motavalli, 40).

Given the degree of regulation and the immense expectations placed on new vehicles, American auto manufacturers evidently emerged from the 1970s feeling under siege. Each industry leader forecasted expensive shifts that would increase vehicle prices and put American laborers out of work as manufacturers were forced to move manufacturing operations abroad to contain costs. In fact, some openly speculated about the fate of the American automobile industry in the 21st century. Later they would apply their considerable creativity to extending the American tradition of car-making into the next century. Nevertheless, American manufacturers obviously chose to direct this creative ability toward circumventing new regulations. In truth, though, air pollution was simply an immediate outcome of burning fossil fuels. The implications and impacts of these emissions would prove even more troublesome to the future of the ICE.

## EMERGING SCIENCE CONNECTS EMISSIONS TO LARGER ECOLOGICAL PROBLEMS

The new appreciation of the environmental impact of the ICE was just the beginning of the problems that would face the brokers of America's high-energy existence. A number of developments during the late 1970s and early 1980s started to move global warming into the political limelight. As a general backdrop, a number of environmental issues became important to the general public and to American politics. One of the first was the problem of urban smog, caused by the release of smoke and sulfur dioxide ($SO_2$) from various human activities. (Volcanoes are a secondary source of $SO_2$ but, of course, do not explain urban smog problems.) Sulfur dioxide reacts with water vapor to form sulfuric acid and other sulfates, which hang in the atmosphere as aerosols. Automobile and industrial emissions also release nitrogen oxides, which combine with various organic compounds to create yet more particulate matter in the atmosphere. Although smog had already been a political issue in the early part of the 20th century, urban pollution continued to be a visible problem in early 1970s, and the American press renewed concern for the issue. Public awareness gave the issue political traction, and legislation, which started with the Clean Air Act of 1963, continued to be passed regularly by Congress, up to the Clean Air Act of 1990.

It was also during the 1970s that concern grew over the environmental issue known as acid rain. The problem was that the various sulfur and nitrogen compounds formed as aerosols in the atmosphere came down with precipitation. This created acidic soil conditions (low pH values), which were not good for plant growth, agriculture, and aquatic life. The Acid Rain Program was an outgrowth of the Clean Air Act of 1990. The program allowed the EPA to head a market-based system in which industries (such as coal-burning power plants) were encouraged to reduce the release of sulfur dioxide and nitrogen oxides by establishing emission quotas and then allowing the companies to buy and sell emission allowances, depending on their economic circumstances and needs.

The growing complications related to air pollution grew when scientists' view of smog was no longer limited to local areas, such as the city of Los Angeles. New computer modeling combined with better understanding of the functioning of various layers of Earth's atmosphere to make clear that something was rapidly depleting the planet's protective ozone layers. In addition, heat was becoming trapped in Earth's atmosphere at an alarming rate, suggesting that the natural greenhouse effect might be increasing. Finally, as discussed previously, by the 1990s scientists concluded that Earth was warming at a pace without historical precedent.

For most scientific observers, the root of each of these environmental problems, as well as others such as acid rain, was the burning of fossil fuels, which released massive amounts of carbon (in the form of carbon monoxide and dioxide) into Earth's atmosphere. The transportation sector alone is responsible for about one-third of our nation's total production of carbon dioxide. And, of course, the ICE is a primary contributor. Not only was smog unpleasant and unhealthy, it may actually be contributing to the ruination of the entire Earth (Gelbspan, 9–13).

Some scientists went even further. They argued that the burning of fossil fuels had broadened humans' environmental impact so severely that a new geological epoch should be named the anthropocene. Chemist Paul Crutzen argued in a 2000 article in *Science* that humans have become a geologic agent comparable to erosion and eruptions, and accordingly "it seems to us more than appropriate to emphasize the central role of mankind in geology and ecology by proposing to use the term 'anthropocene' for the current geological epoch." In this paradigm, the human, particularly its high-energy versions, had become immoral, exploiters of Earth's finite resources. Now, there also appeared to be evidence that our dependence would affect the lives of every human and other living thing on Earth.

## LEARNING LIMITS ON ENERGY IN THE 1970s

Federal discussion about energy use had also begun to change during the 1970s. Stimulated by the 1973 Arab Oil Embargo, American leaders began to perceive our reliance on petroleum (acquired from Middle Eastern countries in increasing amounts by 1970) as a problem. Combining this political reality with the environmental initiatives of the 1970s, a new ethic of resource use and management took shape around the idea of conservation. Although not a culture-wide shift from the era of intense consumption, learning to live within natural limits during the late 1900s took a variety of forms.

Although the embargo had economic implications, it had begun as a political act by OPEC. Therefore, in 1974, the Nixon Administration determined that it needed to be dealt with on a variety of fronts, including, of course, political negotiation. These negotiations, which actually had little to do with petroleum trade, needed to occur between Israel and its Arab neighbors, between the United States and its allies, and between the oil-consuming nations and the Arab oil exporters. Convincing the Arab exporters that negotiations would not begin while the embargo was still in effect, the Nixon Administration leveraged the restoration of production in March 1974. Although the political contentions grew more complex in ensuing decades, the primary impact of the embargo came through the residual effects it had on American ideas of energy.

Now, a panicked public expected action. Richard M. Nixon, by this point increasingly embattled over the growing problem of the Watergate scandal but having nonetheless been reelected in 1972, appeared before Americans on November 7, 1973 to declare an "energy emergency." He spoke of temporary supply problems:

We are heading toward the most acute shortages of energy since World War II. . . . In the short run, this course means that we must use less energy—that means less heat, less electricity, less gasoline. In the long run, it means that we must develop new sources of energy which will give us the capacity to meet our needs without relying on any foreign nation.

The immediate shortage will affect the lives of each and every one of us. In our factories, our cars, our homes, our offices, we will have to use less fuel than we are accustomed to using. . . .

This does not mean that we are going to run out of gasoline or that air travel will stop or that we will freeze in our homes or offices any place in America. The fuel crisis need not mean genuine suffering for any Americans. But it will require some sacrifice by all Americans.

In his speech, Nixon went on to introduce "Project Independence," which, he asserted, "in the spirit of Apollo, with the determination of the Manhattan Project, [would] . . . by the end of this decade" help the nation to develop "the potential to meet our own energy needs without depending on any foreign energy source."

In reality, Nixon's energy czar, William Simon, took only restrained action. Rationing was repeatedly debated, but Nixon resisted taking this drastic step on the federal level. Although he had rationing stamps printed, they were kept in reserve. In one memo, Nixon's aid Roy Ash speculated that: "In a few months, I suspect, we will look back on the energy crisis somewhat like we now view beef prices—a continuing and routine governmental problem—but not a Presidential crisis." Nixon's notes on the document read "absolutely right" and, overall, his actions bore out this approach. He refused to be the president who burst the American high of energy decadence.

Of course, any argument for a conservation ethic to govern American consumers' use of energy was a radical departure from the postwar American urge to resist limits and to flaunt the nation's decadent standard of living. Although this ethical shift did not take over the minds of all Americans in the 1970s, a large segment of the population began to consider an alternative paradigm of accounting for our energy use and needs. The national morals were shifting. They became interested in energy-saving technologies, such as insulation materials and low wattage light bulbs, as well as limits on driving speeds that might increase engine efficiency. As a product of the 1970s crisis, some Americans were even ready and willing to consider less convenient ideas of power generation such as alternative fuels.

One conduit for such research would be the Department of Energy (DOE) that President Jimmy Carter created at the cabinet level. Similar crisis of energy supplies might be avoided, it was thought, if in the future one agency administered strategic planning of energy use and development. The DOE's task was "to create and administer a comprehensive and balanced national energy plan by coordinating the energy functions of the federal government." (DOE) The DOE undertook responsibility for long-term, high-risk research and development of energy technology, federal power marketing, energy conservation, the nuclear weapons program, energy regulatory programs, and a central energy data collection and analysis program.

Just as with any government agency, however, the mandate and funding varies with each presidential election. During the next few decades, the DOE moved away from energy development and regulation toward nuclear weapons research, development, and production. Since the end of the Cold War, the department has focused on environmental cleanup of the nuclear weapons complex, nonproliferation and stewardship of the nuclear stockpile, as well as some initiatives intended to popularize energy efficiency and conservation. As the crisis faded, so too did the political will to strategically plan the nation's energy future.

Although the DOE did not necessarily present Americans with a rationale for less polluting fuels, new ethical perspectives moved through the public after the 1960s and created a steady appreciation for renewable energy. Interest in these sustainable methods as well as in conservation helped spur the public movement in the late 1960s that became known as modern environmentalism.

## APPLYING THE ENVIRONMENTAL PARADIGM TO ENERGY

It did not take long for this new way of viewing the human condition to focus itself on the ethics behind Americans' high-energy lifestyle. The thinker most often given credit for making this transition in thought is E. F. Schumacher, a British economist who, beginning in 1973, wrote a series of essays titled *Small Is Beautiful*. Another title, *Small Is Beautiful: Economics as if People Mattered*, became a best-seller. Like all the books in the series, this one emphasized the need to consider a different view of progress than the expansive, energy-intensive American approach. Building from the idea of limits that the embargo had reinforced, Schumacher emphasized a philosophy of "enoughness," in which Americans designed their desires around basic human needs and a limited, appropriate use of technology. Later, this approach was termed "Buddhist Economics."

Schumacher particularly faults the conventional economic thinking that failed to consider sustainability and, instead, emphasized growth at all costs and a basic trust in the idea that "bigger is better." The key, he argues, was in the conception of new technologies—when inventors and engineers were literally choosing why they pursued an innovation. He writes:

Strange to say, technology, although of course the product of man, tends to develop by its own laws and principles, and these are very different from those of human nature or of living nature in general. Nature always, so to speak, knows where and when to stop. Greater even than the mystery of natural growth is the mystery of the natural cessation of growth. There is measure in all natural things—in their size, speed, or violence. As a result, the system of nature, of which man is a part, tends to be self-balancing, self-adjusting, self-cleansing. Not so with technology, or perhaps I should say: not so with man dominated by technology and specialisation. Technology recognises no self-limiting principle—in terms, for instance, of size, speed, or violence. It therefore does not possess the virtues of being self-balancing, self-adjusting, and self-cleans-mg. In the subtle system of nature, technology, and in particular the super-technology of the modern world, acts like a foreign body, and there are now numerous signs of rejection.

Suddenly, if not altogether surprisingly, the modern world, shaped by modern technology, finds itself involved in three crises simultaneously. First, human nature revolts against inhuman technological, organisational, and political patterns, which it experiences as suffocating and debilitating; second, the living environment which supports human life aches and groans and gives signs of partial breakdown; and, third, it is clear to anyone fully knowledgeable in the subject matter that the inroads being made into the world's non-renewable resources, particularly those of fossil fuels, are such that serious bottlenecks and virtual exhaustion loom ahead in the quite foreseeable future.

Any one of these three crises or illnesses can turn out to be deadly. I do not know which of the three is the most likely to be the direct cause of collapse. What is quite clear is that a way of life that bases itself on materialism, i.e., on permanent, limitless expansionism in a finite environment, cannot last long, and that its life expectation is the shorter the more successfully it pursues its expansionist objectives. (Schumacher)

Although Schumacher's points may have been extreme, they presented a new paradigm of restraint in energy management that appealed to some intellectuals. It was organized by a new ethic of restraint that was tailor-made for the findings of climate science.

One of the most noticeable spokespeople of this alternative energy paradigm was economist Amory Lovins who published an article entitled "Soft Energy Paths" in *Foreign Affairs* in 1976. In his subsequent book, Lovins contrasted the "hard energy path," as forecast at that time by most electrical utilities, and the "soft energy path," as advocated by Lovins and other utility critics. He writes:

The energy problem, according to conventional wisdom, is how to increase energy supplies . . . to meet projected demands. . . . But how much energy we use to accomplish our social goals could instead be considered a measure less of our success than of our failure. . . . [A] soft [energy] path simultaneously offers jobs for the unemployed, capital for businesspeople, environmental protection for conservationists, enhanced national security for the military, opportunities for small business to innovate and for big business to recycle itself, exciting technologies for the secular, a rebirth of spiritual values for the religious, traditional virtues for the old, radical reforms for the young, world order and equity for globalists, energy independence for isolationists. . . . Thus, though present policy is consistent with the perceived short-term interests of a few powerful institutions, a soft path is consistent with far more strands of convergent social change at the grass roots. (Lovins, 102)

Lovins's ideas moved among intellectuals but found immediate acceptance with neither political leaders nor the general public. The shift, though, seemed to arrive in the form of President Carter.

With additional instability in the Middle East by the late 1970s, Carter elected to take the ethic of energy conservation directly to the American people (Horowitz, 20–5). Tying overconsumption to gluttony and immorality, Carter attempted to steer the nation toward a future of energy conservation and independence. In a 1977 speech, Carter urged the nation:

Tonight I want to have an unpleasant talk with you about a problem unprecedented in our history. With the exception of preventing war, this is the greatest challenge our country will face during our lifetimes. The energy crisis has not yet overwhelmed us, but it will if we do not act quickly.

It is a problem we will not solve in the next few years, and it is likely to get progressively worse through the rest of this century.

We must not be selfish or timid if we hope to have a decent world for our children and grandchildren.

We simply must balance our demand for energy with our rapidly shrinking resources. By acting now, we can control our future instead of letting the future control us. . . .

Our decision about energy will test the character of the American people and the ability of the President and the Congress to govern. This difficult effort will be the "moral equivalent of war"—except that we will be uniting our efforts to build and not destroy.

In a risky political move, Carter attempted to steer Americans down a path less trodden—in fact, a path requiring severe difficulty and radical social and cultural transition. It was a path of resource management inspired by the concept of restraint and conservation. It was a lonely argument.

In 1979, Carter took the moral argument one step further in the speech now referred to as his "malaise" speech. Following the release of his book, *Limits of Power,* political historian Andrew Bacevich was interviewed by PBS host Bill Moyers in 2008. Here is what he had to say about Carter's speech and the role of following presidents:

Well, this is the so-called Malaise Speech, even though he never used the word "malaise" in the text to the address. It's a very powerful speech, I think, because President Carter says in that speech, oil, our dependence on oil, poses a looming threat to the country. If we act now, we may be able to fix this problem. If we don't act now, we're headed down a path in which not only will we become increasingly dependent upon foreign oil, but we will have opted for a false model of freedom. A freedom of materialism, a freedom of self-indulgence, a freedom of collective recklessness. And what the President was saying at the time was, we need to think about what we mean by freedom. We need to choose a definition of freedom which is anchored in truth, and the way to manifest that choice, is by addressing our energy problem.

He had a profound understanding of the dilemma facing the country in the post Vietnam period. And of course, he was completely hooted, derided, disregarded.

Moyers then asked Bacevich how Presidents Reagan and Bush who followed Carter responded to this challenge. Bacevich is careful to point out that this "is not a Republican thing, or a Democratic thing, all presidents, all administrations are committed to that proposition [that] . . . the American way of life is not up for negotiation." He continues:

What I would invite them to consider is that, if you want to preserve that which you value most in the American way of life, and of course you need to ask yourself, what is it you value most. That if you want to preserve that which you value most in the American way of life, then we need to change the American way of life. We need to modify that which may be peripheral, in order to preserve that which is at the center of what we value. (Moyers)

Although ensuing presidents backed off from Carter's moral and practical view of American consumption, the connection he made to energy supplies received important validation from the evolving scientific understanding of pollution and climate change.

Growing from a similar worldview after the 1970s, some members of the scientific community also began to make a connection between American patterns of resource use and the health of Earth. Through international

NGOs, this interest was soon turned to global environmental issues, including chlorofluorocarbons (CFCs) air pollution, overfishing, nuclear testing, and, ultimately, global warming. Organizations, particularly stemming from the United Nations, that are discussed in chapters 5 and 6 functioned to move ideas and concepts directly into policy and concrete initiatives around the globe. Suddenly, as the 21st century began, another ripple of morality was added to the issue of climate change: a nation, such as the United States, that seemed to have been a significant contributor to climate change refused to work with the international community to help repair the problem. To many international observers, the U.S. refusal was unconscionable and immoral.

## LOOKING FOR ENERGY ALTERNATIVES

After the 1970s, some policy initiatives specifically focused on the suspected causes of climate change, particularly on the use of hydrocarbons for fuel. Although some policies had by the 1970s begun to recognize pollution and other implications of the use of fossil fuels to create energy, regulations forced energy markets to reflect neither the full environmental nor economic costs of energy production, including potential implications for altering climate. Policy historian Richard Andrews writes that the 1973 embargo initiated three types of policy change related to energy: first, an emphasis on tapping domestic supplies or energy; second, a new recognition that energy conservation was an essential element of any solution; third, electric utility companies were forced to accept and pay fair wholesale rates for electricity created by any producer.

The Public Utilities Regulatory Policy Act of 1978 opened the electric grid to independent producers, including that generated from renewable sources. Eventually, the Energy Policy Act of 1992 expanded these possibilities nationally by allowing both the utilities and other producers to operate wholesale generating plants outside the utility's distribution region. Such initial steps often did not have the intended outcome. Andrews writes that "in effect it thus severed power generation from the 'natural monopoly' of electric transmission and distribution" (Andrews, 301–2).

Carter and others had put the moral imperative for change in front of the American people at the end of the 1970s. Animating the health of the Earth, much as a species of tree or animal that is impacted by human activity, many environmentalists spent the closing decades of the 20th century expanding the scale of our ability to consider the broader implications of our life-style. Most significant, after scientists of the 1980s presented findings that suggested a connection between the burning of fossil fuels and the rise in temperatures on Earth, Americans could begin to appreciate the implications of our trends in lifestyle. In short, more Americans than ever before came to see their everyday tendencies as having moral implications for the planet.

If significant change in energy use was to occur, however, personal transportation would be one of the most difficult alterations for Americans to accept. The dynamics of these new scientific understandings and the ethic of environmentalism altered personal transportation in small pockets as the 20th century closed. The rise of environmental concerns focused in California in the late 20th century and, therefore, it is not surprising that so did the development of electric vehicles. CARB (California Air Resources Board) helped to stimulate CALSTART, a state-funded nonprofit consortium that functioned as the technical incubator for America's efforts to develop alternative-fuel automobiles during the 1990s. Focusing its efforts on the project that became known as the EV (electric vehicle), this consortium faced auto manufacturers' onslaught almost single-handedly. Maintaining the technology during the mid-1900s, however, had been carried out by a variety of independent developers.

Absent of governmental support and despite the contrary efforts of larger manufacturers after World War II, independent manufacturers continued to experiment with creating an electric vehicle that could be operated cheaply and travel farther on a charge. The problems were similar to those faced by Edison and earlier tinkerers: reducing battery weight and increasing range of travel. Some of these companies were already in the auto business, including Kish Industries of Lansing, Michigan, a tooling supplier. In 1961, it advertised an electric vehicle with a clear, bubble roof known as the Nu-Klea Starlite. Priced at $3,950 without a radio or a heater, the car's mailing advertisements promised "a well designed body and chassis using lead acid batteries to supply the motive energy, a serviceable range of 40 miles with speeds on the order of 40 miles an hour." By 1965, another letter from Nu-Klea told a different story: "We did a great deal of work on the electric car and spent a large amount of money to complete it, then ran out of funds, so it has been temporarily shelved" (Motavalli, 40). The Nu-Klea was not heard from again.

In 1976, the U.S. Congress passed legislation supporting the research of electric and hybrid vehicles. Focused around a demonstration program of 7,500 vehicles, the legislation was resisted by government and industry from the start. Battery technology was considered to be so lacking that even the demonstration fleet was unlikely. Developing this specific technology was the emphasis of the legislation in its final rendition. Historian David Kirsch writes that this contributed significantly to the initiative's failure. "Rather than considering the electric vehicle as part of the automotive transportation system and not necessarily a direct competitor of the gasoline car, the 1976 act sponsored a series of potentially valuable drop-in innovations." Such innovations would allow electric technology to catch up to gasoline, writes Kirsch. However, "given that the internal combustion engine had a sixty-year head start, the federal program was doomed to fail" (Kirsch, 205).

Various efforts to create electric vehicles with mass application followed, but nothing that caught on with the American public. In fact, in one of the great moral rejections in energy history, most Americans spent the 1990s purchasing larger and heavier vehicles than ever before: SUVs and pickup trucks became the most popular vehicles on American roads. It appears, however, this tendency was a residual holdover from previous eras in our high-energy existence; genuine cultural and social changes were indeed occurring in the very fabric of American lifestyles. Rising gas prices and the failure of the American automobile industry, ultimately, appear to have helped to alter American vehicle preferences.

## CLIMATE CHANGE AND GREEN CULTURE

Although the concept began in scientific circles, the emphasis on the moral implications of climate change was brought to the American people through popular culture. This message proved to be one of the first conveyed through the mechanisms of green culture, a growing public awareness of basic environmental concepts. Green culture is a product of "soft" environmentalism. Although many Americans who called themselves environmentalists would not change basic consumptive patterns such as the size of their vehicles, upper-middle class consumers and particularly their children possessed an awareness of environmental issues that translated into a passive style of concern. They would often emphasize recycling, animal stewardship, or gardening and celebrate Earth Day. As this ethic or interest moved into popular culture in the 1990s, corporations, films, and television programs began to exploit this green consciousness, sometimes falsely (greenwashing) and other times authentically. By luck of timing primarily, global warming was one of the first large-scale environmental concerns to emerge in cultural soil made ready by green marketing.

In the case of climate change, the first well-known example was a feature film, titled *The Day After Tomorrow* (2004). Similar to many science fiction films that seize on a remote possibility and use the genre to imagine if it came to pass, the big-budget feature film is loosely based on an unlikely scenario related to climate change: the theory of "abrupt climate change." In the film, global warming has resulted in massive shifts in Earth's basic organization: primarily, the Gulf Stream (part of the Atlantic thermohaline circulation) has shut down. This causes the North Atlantic region to cool while heat builds up in the tropics. The result is a severe storm, the likes of which has never before been seen, and a dramatic change in the global climate. As the scenario spins forward, expansive portions of Earth are suddenly plunged into an ice age. In the Untied States, for instance, surviving Americans flee South to Mexico in order to take shelter south of the equator. A bit lost in the wake of the attacks of 2001 and the seemingly impossible scale of the devastation's connection with climate change, *The Day After Tomorrow* did not register fully with the viewing public. Science fiction writer Michael Crichton contributed to this discourse with the novel *State of Fear* in 2004, a cautionary tale about believing all of what scientists tell them. This discourse over an issue of science set the stage, however, for a different type of film, a pseudo-documentary made by Davis Guggenheim and Laurie David.

*Inconvenient Truth* grew from a slideshow presentation about the problem of global warming that was being presented around the country by former Senator and Vice President Al Gore. While in undergraduate at Harvard University, Gore had come into contact with the scientific research of Roger Revelle. Once Gore had been elected to Congress, he began giving these ideas a national platform. After the contested presidential election of 2000, former Vice President Gore devoted himself to the cause of climate change. Gore and film producer David created *An Inconvenient Truth* to serve as a counterbalance to the misinformation about global warming that they felt was being sent to Americans through popular and political culture. By designing a film that would receive a large release and teaming it with other outlets, including a book authored by Gore, Gore and David brought new energy and awareness to the issue. In the film, Gore conveyed an urgency heard rarely by the American public and through the book and Web elements, concerned viewers were given suggestions about how they can create positive action. For instance, in the book *An Inconvenient Truth,* Gore writes:

Although it is true that politics at times must play a crucial role in solving this problem, this is the kind of challenge that ought to completely transcend partisanship. So whether you are a Democrat or a Republican . . . I very much hope that you will sense that my goal is to share with you both my passion for the Earth and my deep sense of concern for its fate. It is impossible to feel one without the other when you know all the facts. (Gore, 10)

Could a former Democratic senator, vice president, and candidate for president achieve nonpartisanship on this important issue? No. And this has greatly limited some of the public's acceptance of *An Inconvenient Truth;*

however, Gore continued in his Introduction to raise the issue to the level of morality. He writes:

I also want to convey my strong feeling that what we are facing is not just a cause for alarm, it is paradoxically also a cause for hope. As many know, the Chinese expression for "crisis" consists of two characters side by side. The first is the symbol for "danger," the second the symbol for "opportunity." The climate crisis is, indeed, extremely dangerous. In fact it is a true planetary emergency. (Gore, 10)

Gore describes the disruption of some of Earth's basic natural processes. In the process, he does not quibble with causality; humans' contribution to climate change is the culprit. He continues by attempting to put these changes in geological context:

Global warming, along with the cutting and burning of forests and other critical habitats, is causing the loss of living species at a level comparable to the extinction event that wiped out the dinosaurs 65 million years ago. That event was believed to have been caused by a giant asteroid. This time it is not an asteroid colliding with the Earth and wreaking havoc; it is us. (Gore, 10)

For Gore, the connection to morality was also based on the need for quick action. To elect to do nothing invites cataclysm. He brings to bear the "exceptionally strong" scientific consensus represented by the work of the Intergovernmental Panel on Climate Change (IPCC), noting its insistence that the world's nations must cooperate to combat global warming. Yet in particular, he bypasses political leaders to close with a call for grassroots action. He writes:

So the message is unmistakably clear. This crisis means "danger"! Why do our leaders seem not to hear such a clear warning? Is it simply that it is inconvenient for them to hear the truth? If the truth is unwelcome, it may seem easier just to ignore it. But we know from bitter experience that the consequences of doing so can be dire . . .

Today, we are hearing and seeing dire warnings of the worst potential catastrophe in the history of human civilization: a global climate crisis that is deepening and rapidly becoming more dangerous than anything we have ever faced. . . . As Martin Luther King Jr. said in a speech not long before his assassination: "We are now faced with the fact, my friends, that tomorrow is today. We are confronted with the fierce urgency of now. In this unfolding conundrum of life and history, there is such a thing as being too late." (Gore, 10–11)

*Inconvenient Truth* took great care not to make the debate only a moral one. Shifting to new energy sources and systems was depicted as a great engine for economic growth and development. Using the film as a mobilizing device,

Gore and David created Web-based modules to give individuals ideas on how to actively combat climate change in their communities. In addition, massive workshops took place to train a corps of inspired Americans to give talks, show the film, and generally fuel community awareness about climate change at the grassroots level. Still, the clear emphasis was on doing the "right" thing for Earth. Gore continues:

But there's something even more precious to be gained if we do the right thing. The climate crisis also offers us the chance to experience what very few generations in history have had the privilege of knowing: *a generational mission*; the exhilaration of a compelling *moral purpose*; a shared and unifying *cause*; the thrill of being forced by circumstances to put aside the pettiness and conflict that so often stifle the restless human need for transcendence; *the opportunity to rise*.

When we do rise, it will fill our spirits and bind us together. . . . When we rise, we will experience an epiphany as we discover that this crisis is not really about politics at all. It is a moral and spiritual challenge. (Gore, 11)

The film's reception was remarkable. The film's astonishing popularity, its Emmy award in documentary film making, and massive sales was topped only by human civilization's greatest stamp of moral correctness: in 2007 the IPCC and Gore, for his work with *An Inconvenient Truth,* were awarded the Nobel Peace Prize.

In their declaration of the award, the Nobel committee said:

The Norwegian Nobel Committee has decided that the Nobel Peace Prize for 2007 is to be shared, in two equal parts, between the Intergovernmental Panel on Climate Change (IPCC) and Albert Arnold (Al) Gore Jr. for their efforts to build up and disseminate greater knowledge about man-made climate change, and to lay the foundations for the measures that are needed to counteract such change.

Indications of changes in the earth's future climate must be treated with the utmost seriousness, and with the precautionary principle uppermost in our minds. Extensive climate changes may alter and threaten the living conditions of much of mankind. They may induce large-scale migration and lead to greater competition for the earth's resources. Such changes will place particularly heavy burdens on the world's most vulnerable countries. There may be increased danger of violent conflicts and wars, within and between states.

Through the scientific reports it has issued over the past two decades, the IPCC has created an ever-broader informed consensus about the connection between human activities and global warming. Thousands of scientists and officials from over one hundred countries have collaborated to achieve greater certainty as to the scale of the warming. Whereas in the 1980s global warming seemed to be merely an interesting hypothesis, the 1990s produced firmer evidence in its support. In the last few years, the connections have become even clearer and the consequences still more apparent.

Al Gore has for a long time been one of the world's leading environmentalist politicians. He became aware at an early stage of the climatic challenges the world is facing. His strong commitment, reflected in political activity, lectures, films and books, has strengthened the struggle against climate change. He is probably the single individual who has done most to create greater worldwide understanding of the measures that need to be adopted.

By awarding the Nobel Peace Prize for 2007 to the IPCC and Al Gore, the Norwegian Nobel Committee is seeking to contribute to a sharper focus on the processes and decisions that appear to be necessary to protect the world's future climate, and thereby to reduce the threat to the security of mankind. Action is necessary now, before climate change moves beyond man's control. (Nobel)

Few greater stamps of integrity and authenticity exist in contemporary life than the Nobel Peace Prize. The issue of global warming had clearly transcended any other environmental issue in global awareness. As the first overt shot toward compelling the climate debate into a morality *An Inconvenient Truth* set the battle lines of the conflict that was to follow. To those inspired by the film, to be against any action to combat climate change completely defied logic and scientific understanding.

## OF MORAL-SUASION AND NAYSAYERS

Those who dispute the idea of humans' role in climate change derive their viewpoints from a variety of rationales; for some, morality was a secondary issue. Instead, they emphasize faulty science. Freeman Dyson, the 80-year-old scientific contrarian at Princeton University has been disputing the concept for the past five years, calling the effects "grossly exaggerated." In a 2007 interview with Salon.com, he emphasized his point further by saying: "the fact that the climate is getting warmer doesn't scare me at all." In other writings, he said that climate change had become an obsession of a type of faith in America known as environmentalism.

The primary reason for this critique, however, is what Dyson calls "lousy science." In early 2009, he stated that he considers the attention given to this issue as immoral given the greater, more serious—and verifiable—dangers that humans face. Other critics have picked up on Dyson's demand for better scientific proof. William Chameides, dean of the Nicholas School of the Environment and Earth Sciences at Duke University, told the *New York Times,* "I don't think it's time to panic," but contends that, because of global warming, "more sea-level rise is inevitable and will displace millions; melting high-altitude glaciers will threaten the food supplies for perhaps a billion or more; and ocean acidification could undermine the food supply of another billion or so." In addition, the existing science is not conclusive. Although the evidence may have the ring of potential plausibility, it is neither truth nor fact. And, argue critics, we are proposing to alter our very existence because of it.

Many scientists, particularly Dyson, have especially grown troubled by the proponents' use of a moral argument to justify action. This Dyson classifies as presumptuous. In many of his public appearances, he warns audiences of the litany of false predictions that litter scientific experimentation. In these moments, Dyson said to the *New York Times,* for scientists, "I think it's more a matter of judgment than knowledge."

To Dyson and others, basing action on incomplete science was irresponsible and, to a large degree, immoral. For many other naysayers, though, the immorality of acting to stop climate change was based on very basic matters of faith and spirituality. For instance, in many Christian faiths resources are put here by God for human use (see, for instance, Genesis, 1:28). To restrain use of resources and therefore human development is immoral. Combine this perspective with the patriotic belief that Americans restrain their development at the peril of the nation's perpetuity, and it becomes clear to see the extremes to which some critics of global warming would go. In a nation prioritizing corporate and industrial growth, the idea of such radical changes was often categorized as soft or even immoral.

In the U.S. Senate, for instance, many pro-development Republicans refused to allow the United States to become involved in the global discourse on locating a political remedy to climate change. By far the most outspoken opponent of the effort to develop such policy solutions was U.S. Sen. James M. Inhofe (R-Okla.), particularly when he served as chairman of the Committee on Environment and Public Works. For instance, on July 28, 2003, Inhofe made the remarks about "The Science of Climate Change" on the floor of the Senate in which he sought to discuss the morality of using science as a basis for policy:

That's why I established three guiding principles for all committee work: it should rely on the most objective science; it should consider costs on businesses and consumers; and the bureaucracy should serve, not rule, the people.

Without these principles, we cannot make effective public policy decisions. They are necessary to both improve the environment and encourage economic growth and prosperity.

One very critical element to our success as policymakers is how we use science. That is especially true for environmental policy, which relies very heavily on science. I have insisted that federal agencies use the best, non-political science to drive decision-making. Strangely, I have been harshly criticized for taking this stance. To the environmental extremists, my insistence on sound science is outrageous.

For them, a "pro-environment" philosophy can only mean top-down, command-and-control rules dictated by bureaucrats. Science is irrelevant-instead, for extremists, politics and power are the motivating forces for making public policy.

But if the relationship between public policy and science is distorted for political ends, the result is flawed policy that hurts the environment, the economy, and the people we serve.

Sadly that's true of the current debate over many environmental issues. Too often emotion, stoked by irresponsible rhetoric, rather than facts based on objective science, shapes the contours of environmental policy. (Inhofe)

Inhofe went on to use this moment to specifically emphasize global warming:

Today, even saying there is scientific disagreement over global warming is itself controversial. But anyone who pays even cursory attention to the issue understands that scientists vigorously disagree over whether human activities are responsible for global warming, or whether those activities will precipitate natural disasters.

I would submit, furthermore, that not only is there a debate, but the debate is shifting away from those who subscribe to global warming alarmism. After studying the issue over the last several years, I believe that the balance of the evidence offers strong proof that natural variability is the overwhelming factor influencing climate.

It's also important to question whether global warming is even a problem for human existence. Thus far no one has seriously demonstrated any scientific proof that increased global temperatures would lead to the catastrophes predicted by alarmists. In fact, it appears that just the opposite is true: that increases in global temperatures may have a beneficial effect on how we live our lives.

For these reasons I would like to discuss an important body of scientific research that refutes the anthropogenic theory of catastrophic global warming. I believe this research offers compelling proof that human activities have little impact on climate. (Inhofe)

Therefore, as forces gathered in favor of political action aimed at mitigating global warming, an equal and opposite set of forces built against such action. While the supporters of the consensus view gained allies such as Gore, the global warming skeptics gained allies such as Inhofe. He claimed that this evidence against the consensus view had been gathered by "the nation's top climate scientists," citing as support for this the Heidelberg Appeal and the Oregon Petition (see chapter 5). In a familiar trope of contemporary environmentalism, each side of the argument appeared to have the support of scientific findings. This aspect of the global warming debate would prove particularly confounding to the American public.

In this political discussion, science became ammunition. That often meant that politicians, even the U.S. Congress, seemed to be interfering with scientific controversies. One such occasion concerned the 1998 article by Mann, Bradley, and Hughes (MBH), which had contained the "hockey stick graph" displaying the average global temperatures of the last thousand years (and showing a sudden increase during the last hundred). The graph, which was used in the third IPCC report, received significant scientific criticism in an article by Stephen McIntyre and Ross McKitrick in 2003. This was followed by a somewhat more politicized report presented to Congress in 2006.

Alarmed that public policy might be based on incorrect science, U.S. Representatives Joseph Barton and Edward Whitfield asked a team of statisticians, led by Edward Wegman, to review the MBH work. The so-called Wegman report suggested that MBH did not apply the best statistical methods to their data set and relied on temperature measurement proxies that were sometimes unreliable (in particular, the bristlecone pine tree ring series). Despite these criticisms, and despite the fact that the 1998 MBH article, like all new research, contained errors and uncertainties, the results were largely confirmed in further work by the original authors and by other scientists. Certainly, though, this situation and innumerable others provided significant opportunity for public confusion on the issue of global warming.

Finally, the pro-development stance was also stimulated by organizations such as the CATO Institute, which prioritizes free markets. Publishing its own books on the topic of climate change, the Institute produced Patrick J. Michaels and Robert C. Balling, Jr.'s *Climate of Extremes: Global Warming Science They Don't Want You to Know* (2009) as well as Michaels' *Meltdown: The Predictable Distortion of Global Warming by Scientists, Politicians, and the Media* (2005).

Working with the Bush Administration, Inhofe and other naysayers and deniers were able to categorize the global warming faction as "alarmists" in much the same fashion that nonconformists were called "red" or "commies" in previous decades. The difference, of course, was the basic ethic behind the effort to craft policies to help put off climate change was not devious or un-American. History has shown that many of the naysayers were more a product of fearing change and uncertainty than of actual scientific findings.

## JAMES HANSEN, GEORGE W. BUSH, AND THE CONTROL OF SCIENCE

The frontlines of the cultural battle over global warming moved swiftly into politics when President George W. Bush took office in 2001. Publicly, he was skeptical of the alarm and panic over global warming; privately, he led a profound effort by the executive branch to stifle scientific findings by other branches of the government. In *Censoring Science,* Mark Bowen writes:

After less than two months in offices, the new president, George W. Bush, had announced that he would abandon a campaign promise to regulate carbon dioxide from coal-burning power plants . . . and then swiftly pulled out of the Kyoto Protocol. . . . As Christine Todd Whitman, then the administrator of the Environmental Protection Agency, later put it, this "was the equivalent to 'flipping the bird,' frankly, to the rest of the world." (Bowen, 1)

In particular, with the support of large energy interests, the Bush administration focused significant effort in tempering and even squelching scientists from NASA who worked with the modeling and satellite imagery related to the government's investigation of climate change. In 2006, Dr. James E. Hansen, director of NASA's Goddard Institute for Space Studies, went public with accusations that the executive branch was prohibiting him from bringing his findings on global warming to the public. During the 1980s, Hansen's group had predicted that the impact of $CO_2$ emissions on global warming would become clear by the end of the 20th century. Although Hansen has complained of being limited by both Democratic and Republican administrations whenever his conclusions contradicted the nation's energy policy, his experiences with the Bush Administration were perhaps the worst of his career.

The Bush Administration sharply reduced NASA's earth science budget and appointed its own people to key NASA positions. The appointees often had limited scientific expertise but maintained ideological positions that were acceptable to the administration. There are a number of documented cases in which senior NASA managers (such as George Deutsch) and officials in the White House (such as Philip Cooney) rewrote scientific reports to the government such that they would highlight scientific uncertainties and minimize the danger presented by global warming. On one occasion in late 2005, Jim Hansen was stopped by NASA officials from conducting an interview with National Public Radio on the grounds that the network was too liberal (Bowen, 24).

In 2006, Hansen told journalists that the President's operatives had ordered the public affairs staff to review his coming lectures, papers, postings on the Goddard Web site and requests for interviews from journalists. Dr. Hansen said he would ignore the restrictions. "They feel their job is to be this censor of information going out to the public," he told the *New York Times.* He released a 2004 e-mail from the White House that instructed him that "The White House (is) now reviewing all climate related press releases," In reports eventually released to the public, the reports of Hansen and other NASA scientists were shown to have endured serious revision at the hand of information officers working with the White House (Bowen, 35).

Hansen, one of the world's leading climate scientists, went on to endure relentless criticism from thinkers who believed it was immoral to take action to try to stop climate change. Although the Bush Administration ultimately softened its view and began to accept that some action on climate change had to take place, the issue of controlling the scientific findings of those such as Hansen has been allowed to fade to the past—an indication of anxiety over the changes needed to combat climate change. Hansen, however, has not gone away. Before President Barack Obama's inauguration in January 2009, Hansen and his wife Anniek sent the following open letter to the new leader:

There is a profound disconnect between actions that policy circles are considering and what the science demands for preservation of the planet. A stark scientific conclusion, that we must reduce greenhouse gases below present amounts to preserve nature and humanity, has become clear to the relevant experts. . . . Science and policy cannot be divorced. It is still feasible to avert climate disasters, but only if policies are consistent with what science indicates to be required. Our three recommendations derive from the science, including logical inferences based on empirical information about the effectiveness or ineffectiveness of specific past policy approaches.

1. Moratorium and phase-out of coal plants that do not capture and store $CO_2$. . . .
2. Rising price on carbon emissions via a "carbon tax and 100 percent dividend.". . .
3. Urgent R&D on fourth generation nuclear power with international cooperation.

Summary
An urgent geophysical fact has become clear. Burning all the fossil fuels will destroy the planet we know, Creation, the planet of stable climate in which civilization developed.

Of course it is unfair that everyone is looking to Barack to solve this problem (and other problems!), but they are. He alone has a fleeting opportunity to instigate fundamental change, and the ability to explain the need for it to the public. (Hansen)

Although not all scientists agreed with Hansen's findings, NASA's models provided convincing evidence to many that the situation merited immediate attention. In his criticism, Hansen argued that it was immoral for the federal government not to report the information it had found, which could influence the lives of every human.

## CONCLUSION: GLOBAL WARMING BECOMES MAINSTREAM

The 2008 ad summed up how rapidly the debate on global warming was changing: Created by Gore's *Alliance for Climate Protection,* the public relations entity he created to manufacture green culture concerning climate change, featured Newt Gingrich, former House Speaker and unofficial spokesman of the Republican Conservatives, sitting on a small couch—often referred to as a "love seat"—with Nancy Pelosi, current House Speaker and representative of the Democrat's more liberal thinkers, with the Capitol looming large in the background. They confess that they don't "always see eye-to-eye," but, Gingrich states directly into the camera, "we do agree our country must take action to address climate change" (Gingrich ad). In the last year, Gingrich and many other former naysayers altered their discourse to accept the concept of climate change—and often also the need for action—but to debate what role the federal government should take in such action. The ad functions as a symbol of how broadly known the issue of climate change had become and, it seemed, how mainstream had become the consensus for action. A significant portion of this moment came from efforts to overcome one of the naysayers' primary, moral arguments against acting to stop climate change.

This new era had begun with specific efforts by some scientists to explain their findings to religious leaders and by some religious leaders to cast action as an important component of "planetary stewardship." For instance, in an interview with the *New Statesman,* Sir John Houghton, former director general and chief executive of the Meteorological Office (1983–91), chairman or co-chairman of the Scientific Assessment Working Group of the Inter-governmental Panel on Climate Change (1988–2002), and chairman of the Royal Commission on Environmental Pollution (1992–98) was asked how he persuaded evangelical leaders to launch their efforts on behalf of stopping global warming. He responded:

The process began in a meeting I organised in Oxford in 2001, after I happened to meet Calvin DeWitt, a distinguished professor of environmental studies at Wisconsin who also runs the Au Sable Institute, which educates students on the relation between ecology and the Christian faith. I brought along some of the leading IPCC scientists, including Bob Watson. By no means all were Christians. The Bishop of Liverpool talked about theology. And Richard Cizik, the vice-president of the National Association of Evangelicals [in Washington, DC], said he was very impressed with the honesty and the humility of all the scientists. He went away determined to try and help Christians get to grips with the issue. The result was the Evangelical Climate Initiative, launched [in 2006]. (Houghton)

After an often antagonistic era in which the conservative right argued extreme convictions on divisive issues, Houghton's efforts succeeded by approximately 2004 to form a growing group of influential pastors and the heads of some large faith organizations who set a new national policy agenda founded on their understanding of the life of Jesus and his ministry to the poor, the outcast, and the peacemakers. The movement has no single charismatic leader, no institutional center, and no specific goals. It doesn't even have a name. On issues such as global warming, however, this new Christian thinking emphasizes environmentalism as stewardship of Earth and its components. Instead of casting legislating to stop global warming as an immoral blight on humans, such reforms became moral correctives on the exploitation of Earth's resources. Climate change legislation became an act of stewardship.

Earlier, when very few Christians were willing to declare global warming a problem requiring political attention, Richard Cizik, a lobbyist for the National Association of Evangelicals, led a lonely journey. He argued a variety of environmental stances, but became a national figure with his stance in support of creating policies to attempt to stop or slow global warming. He became particularly outspoken after hearing evidence of global warming in 2002. For Cizik, the issue fell into what he called "creation care," a type of environmentalism that did not stem from political agendas or ideology. Instead, he used biblical verses to support his belief that being a good steward of Earth went along with basic tenets of Christianity. These ideas included:

On the Care of Creation: An Evangelical Declaration on the Care of Creation
   *The Earth is the Lord's, and the fulness thereof*—Psalm 24:1
   As followers of Jesus Christ, committed to the full authority of the *Scriptures*, and aware of the ways we have degraded creation, we believe that biblical faith is essential to the solution of our ecological problems.
   Because we worship and honor the Creator, we seek to cherish and care for the creation.
   Because we have sinned, we have failed in our stewardship of creation. Therefore we repent of the way we have polluted, distorted, or destroyed so much of the Creator's work. . . .
   These degradations of creation can be summed up as 1) land degradation; 2) deforestation; 3) species extinction; 4) water degradation; 5) global toxification; 6) the alteration of atmosphere; 7) human and cultural degradation.
   Many of these degradations are signs that we are pressing against the finite limits God has set for creation. With continued population growth, these degradations will become more severe. Our responsibility is not only to bear and nurture children, but to nurture their home on earth. We respect the institution of marriage as the way God has given to insure thoughtful procreation of children and their nurture to the glory of God. (Cizik)

In 2006, Cizik was instrumental in establishing the Evangelical Climate Initiative and helping it to draft a national "Call to Action." His work has helped many Americans to cross an important threshold toward adopting climate change as an issue of importance in their lives. He has provided a new moral framework to help many Americans who were not schooled in science to grasp basics of climate change.

Gingrich and others picked up on the emphasis of Cizik on stewardship. By April 2009, Gingrich had accepted the centrality of the issue of climate change in the political sphere and had attempted to master it. As Congress considered new initiatives that sought to use federal authority to limit emissions, Gingrich squared off against Gore. Gore, the Democrat, compared the bill's significance to civil rights laws, while the Republican dismissed it as "micromanagement" that invited corruption. Gore described the draft bill to curb greenhouse gas emissions and promote renewable energy as "one of the most important pieces of legislation ever introduced in the Congress." He continued, "I believe this legislation has the moral significance equivalent to that of the civil rights legislation of the 1960s and the Marshall Plan of the late 1940s."

Gingrich pointed out two parts of the bill he liked: the development of a smart electric grid to prevent blackouts and make energy transmission more efficient and the promotion of what he called "green coal technology," which involves finding a way to capture and store carbon from coal-fired power plants responsible for half the nation's electricity. However, overall, "It is ridiculous to believe that we are going to eliminate 83 percent of carbon use with current technologies," he added. "This is the strategy imposed in the bill and it is a fantasy. Nothing in this bill leads to the level of breakthrough that you need to reduce carbon not only here at home but also reduce carbon generated by China and India."

For the issue of climate change, the closing decade of the 20th century and the first of the 21st were marked by public division and discussion. Major breakthroughs occurred when leaders such as Al Gore enabled the general public to relate to and appreciate the magnitude of the issue of global warming. This cultural acceptance, particularly joined with religious groups, set the stage for dramatic political action on the issue of climate change at the start of the 21st century. Contest or debate between ways of life and schools of belief made sure, however, that global warming remained a great source of debate.

# 5

# The International Response

Although climate change was not a major international issue during the 1970s, it reached the world political stage by the end of the 1980s. As we saw in chapter 2, during the 1970s, there was considerable scientific uncertainty regarding global temperature trends. In fact, the instrumental record showed lowered temperatures from the 1940s to the 1970s. Although this trend proved to be temporary, this was not before a number of scientists and journalists warned of "a great cooling." In addition to uncertainty regarding temperature trends, data on past climate, as well as the crucial computer models needed to model global climate processes, were still in a state of infancy.

Nevertheless, climate science was helped during the 1970s by rising interest in environmentalism. Chapter 5 recounts this story for the United States, reviewing problems such as urban air pollution and acid rain. Against this background of rising concern about environmental issues, the issue of global warming reached the international political stage during the late 1980s. The global movement for action, however, would have been impossible without tremendous improvements in the existing scientific knowledge. The first improvements concerned data on global temperature trends. After the uncertainty of the 1970s, global temperatures returned to a rising trend, similar to the one noticed earlier in the century. Also, improved climate data and computer models convinced climate scientists of the plausibility that this global warming was caused by human activity. Therefore, during the 1980s, a consensus about global warming took root (see chapter 2). This chapter tells the story of the action that followed on the international stage.

## A GLOBAL CONCERN

Spurred in part by the public interest in the environment during the 1970s, the federal government initiated efforts, albeit limited, to conduct scientific research in climate change. In 1970, it formed the National Oceanic and Atmospheric Administration (NOAA) as a sort of "wet NASA" and, to some degree, climate science benefited from this. Four years later, a report sponsored by the National Academy of Science issued mild warnings about the risks associated with climate change. In 1978, the climatologist Reid Bryson published the influential book *Climates of Hunger,* which noted the impacts of drought on past human civilizations and warned of possible difficulties for our own. (It is interesting to note that Bryson came to embrace the idea that the Earth's average temperature was rising but remained skeptical of the hypothesis that this was caused by human activity.) In 1978, the National Climate Act was passed by Congress and established an office within NOAA to keep an eye on the problem. The new office, the National Climate Program, however, did not initially receive a significant amount of funding and was not able to do much significant research.

In addition, growing interest in environmentalism became an international phenomenon. Truly global issues, including smog, acid rain, and others, provided nations with new opportunities at cooperative legislation. One of the more alarming and speculative problems was "nuclear winter." During the early 1980s, a number of scientists (led by Carl Sagan) warned of a climate catastrophe in the event of a nuclear war. Because a full-fledged nuclear conflict would involve the explosion of hundreds or even thousands of bombs, massive clouds of dust would form over the surface of the Earth. In addition to the physical damage (including fires) and radiation brought by the bombs, the dust clouds would block sunlight. Computer simulations suggested that this, in turn, would likely lead to widespread global cooling and, eventually, the extinction of many plants and animals. Although there continued to be considerable doubt about the predictions offered by the computer simulations, it was clear that even a minimal effort of prudence would require international awareness of the issue.

Another international issue of the 1980s was much more certain scientifically and more important politically: ozone depletion. Because ozone ($O_3$) in the upper atmosphere absorbs ultraviolet radiation, it provides a crucial canopy for many forms of life on Earth. Unfortunately, the concentrations of free-radical catalysts in the Earth's atmosphere had increased greatly

during the 1970s, as a result of the release of large quantities of manufactured compounds such as chlorofluorocarbons (CFCs). The release of chlorine from the CFCs into the atmosphere was disastrous for the ozone, as the
series of chemical reactions initiated by the chlorine and ozone resulted in the
breakdown of the $O_3$ (into $O_2$) and the generation of fresh chlorine (which is
then free to catalyze the breakdown of yet more ozone).

In 1985, the Vienna Convention for the Protection of the Ozone Layer
established a "framework" (little more than a statement of hopes) for defining
international regulations on ozone-depleting substances. Public and political
concern for the issue increased during the 1980s, however, after actual holes in
the ozone were confirmed over the Earth's poles. A convincing argument was
made by scientists and policymakers that not to act on ozone depletion would
lead to numerous health problems (such as skin cancer), brought on by exposure
to ultraviolet radiation, which would eventually cost society substantial sums.
Two years later, in 1987, the Vienna framework was given teeth in the Montreal
Protocol. The protocol specified formal emission restrictions and led 20 nations
(including the United States) to sign on. Climate scientists, politicians, and
members of the general public, who were increasingly concerned about global
warming, saw an inspiring example in the problem of ozone depletion. The
international community had discovered a serious problem associated with
human emissions and then acted to correct it. As a consensus took shape in the
science behind global warming, the call for action increased (Flannery, 213).

In the United States a number of scientific reviews sounded increasingly
strong warnings to policymakers. For example, in 1983, the National Academy
of Sciences (NAS) issued a conservatively worded study on the impact of rising
$CO_2$ levels and the greenhouse effect. Although the authors said that they were
"deeply concerned" about the possibility of unexpected changes, they also felt
that the changes would not be severe and that additional regulation was not
needed. Just a few days after the NAS report, the Environmental Protection
Agency (EPA) released its own report. Although the EPA report confirmed
many of the NAS conclusions, its warnings were stronger, in particular seeing bigger temperature rises in the coming decades. The Reagan Administration, which was strongly opposed to new government regulation of American
industry, suggested that the EPA report was alarmist. Nevertheless, Congressional hearings followed, which brought global warming a bit more into the public eye. Although no new funding for the environmental sciences was provided
for by the Reagan Administration, neither were there any cuts (Weart, 141).

Arguably, the rise of global warming as a national and international
political issue exemplifies the ever greater development of public knowledge
and democracy: simply, the public was much more aware of basic ecological
principles than at any time in the past. By the late 1980s, as the result of

increased awareness of environmental problems, greater reporting in the media, and responsiveness of the political system, global warming was on the cusp of receiving a place in the public consciousness and ready to take off should something happen to demand action.

The year 1988 came as if on cue. First, the heat waves and droughts of the American summer of 1988, especially in the central and eastern states, made a great deal of news in magazines, newspapers, and TV programs. The National Climatic Data Center (of NOAA) estimated that there were about $40 billion in economic losses and from 5,000 to 10,000 deaths. At the start of this difficult summer, in late June, NASA scientist James Hansen gave testimony to Congress in which he said that it was virtually certain that human activity was responsible for the global warming trends and that this might bring more storms, floods, and heat waves in the future. Hansen made the now-famous remark that "It's time to stop waffling . . . and say that the greenhouse effect is here and is affecting our climate now" (Weart, 150). Along with stories about the severe weather, the media also reported on Hansen's remarks and remarks by other scientists and politicians. Polls indicated that, over the next year, public awareness of global warming rose considerably.

The United Kingdom had a particularly strong interest in the issue because, being a relatively small nation geographically, it stood to lose a great deal from climate change. For one thing, its temperate climate was dependent on the Gulf Stream, which was potentially endangered by global warming. In 1988, the conservative Prime Minister Margaret Thatcher (who had a degree in chemistry from Oxford University) became the first major politician to embrace global warming as an important issue. After being briefed on the science of global warming by her adviser Crispin Tickell, Thatcher brought the issue to the attention of the nation's scientists in a major speech to the Royal Society.

## THE INTERGOVERNMENTAL PANEL ON CLIMATE CHANGE

In 1988, rising scientific interest and political concern about global warming gave rise to the foundation of the Intergovernmental Panel on Climate Change (IPCC) by the World Meteorological Organization and the United Nations (UN). The IPCC's task was to organize the work of an international group of scientists to produce periodic reports summarizing the best knowledge on the world's climate in order to help political leaders make policy decisions. It might come as a surprise that the conservative administration of Ronald Reagan was strongly in favor of forming the body. But Reagan's enthusiasm for the IPCC was quite different from Thatcher's. Whereas Thatcher appears to have been genuinely concerned about the possible impact of global warming on the United Kingdom, Reagan felt that the issue was overblown by the scientists and that acting on it would harm the U.S. economy. His administration therefore did not wish scientists to appoint their own committees or to speak on the issue individually. By contrast, the IPCC, as an official governmental channel of discussion, would hopefully flatten out the more extreme opinions of some of the scientists and lead them to somewhat more cautious conclusions (Weart, 152).

The first IPCC report of 1990 concluded that, although emissions resulting from human activities were certainly increasing the atmospheric concentrations of greenhouse gases, the recent warming trend was probably due to a mix of natural and human causes. For the 21st century, the IPCC predicted that, under a business-as-usual scenario (in which human activity continued as before and mitigation efforts were not made), the global mean temperature would increase in a range between 1.5 and 4.5 degrees Celsius. The IPCC also judged that it would take 10 years before it would be clear how much of the temperature increases were due to natural causes and how much were due to human activity. In sum, the first report did not give particularly alarming or surprising news. In the United States, the media and the public took little note and it had only minor political impact.

In contrast to the United States, response to the first IPCC report by Margaret Thatcher's government was supportive and enthusiastic. Speaking at the Second World Climate Conference in Geneva on November 6, 1990, Prime Minister Thatcher strongly supported the IPCC and climate research generally:

The IPCC tells us that, on present trends, the earth will warm up faster than at any time since the last ice age. Weather patterns could change so that what is now wet would become dry, and what is now dry would become wet. Rising seas could threaten the livelihood of that substantial part of the world's population which lives on or near coasts. The character and behaviour of plants would change, some for the better, some for worse. Some species of animals and plants would migrate to different zones or disappear for ever. Forests would die or move. And deserts would advance as green fields retreated. . . .

The IPCC report is very honest about the margins of error. Climate change may be less than predicted. But equally it may occur more quickly than the present computer models suggest. Should this happen it would be doubly disastrous were we to shirk the challenge now. I see the adoption of these policies as a sort of premium on insurance against fire, flood or other disaster. It may be cheaper or more cost-effective to take action now than to wait and find we have to pay much more later. (Thatcher)

In 1990, Thatcher also supported the foundation of the Hadley Centre for Climate Prediction and Research in Exeter. The center was named after the

18th-century scientist George Hadley, who had discovered the Earth's major circulation cells. It was established by Welsh scientist John Houghton, who had been the lead editor for the first IPCC report (and would be for the following two). Interestingly, Houghton was an evangelical Christian who felt that Christians had responsibilities to be stewards of God's Earth. (Houghton later had contact with American Christian leader Richard Cizik; see chapter 4.) Since then, the Hadley Centre has been United Kingdom's principal site for interpreting climate data, understanding the processes underlying climate change, and developing new computer models for use by the world climate science community.

Although the lowest common denominator approach of the IPCC did lead, at first, to relatively conservative conclusions, each successive IPCC report gave clearer, stronger warnings. The consensus on anthropogenic global warming continued to deepen during the late 1990s and early 2000s. As always, with the publishing of one IPCC report, work was immediately begun on the next. The second report of 1995 combined the work of over twice the number of scientists, about 400 in all. The specific conclusions of the second report did not change greatly from the first. Nevertheless, it was clear that the scientist's confidence had increased. One thing that was stronger about the second IPCC report was its conclusion that human influence on global climate was discernible. Progress had been made in distinguishing between natural causes and anthropogenic causes, largely thanks to improved computer simulations. Reflecting both caution and a greater confidence, the IPCC concluded: "Our ability to quantify the human influence on global climate is currently limited because the expected signal is still emerging from the noise of natural climate variability. . . . Nevertheless, the balance of evidence suggests a discernible influence on global climate" (IPCC, 1996).

The second IPCC report forecast that, by the mid-21st century, atmospheric $CO_2$ levels would probably double and the Earth's average surface air temperature would increase somewhere in the range of 1.5 to 4.5 degrees Celsius. The IPCC hesitated to overclaim. A few of the new computer models projected temperature increases for the 21st century that were above the range of 1.5 to 4.5 degrees Celsius, which was given in the first report. The panel decided against quoting these results, however, and stuck with the 3.0 degree Celsius average. This number actually went back to an even earlier report that had been issued in 1979 by the NAS (the Charney Committee). The IPCC scientists decided that it was wiser to stick with the same numbers to avoid giving the impression that their new work was inconsistent with earlier results (Weart, 165).

Between the second and third reports, there were a wide range of scientific and political challenges to the scientific consensus (see chapter 5). Nevertheless, the international consensus on global warming continued to strengthen, as the IPCC report of 2001 made clear by making significantly stronger conclusions. The third report announced that there was now greater warming than anything seen in the last 10,000 years and that there was strong evidence that most of the warming observed in recent decades could be explained by human activities, especially the emission of greenhouse gases. By comparison, natural factors (such as solar activity and the Milankovitch cycles) made small contributions. The IPCC concluded that "in light of new evidence and taking into account the remaining uncertainties, most of the observed warming over the last 50 years is likely to have been due to the increase in greenhouse gas concentrations" (IPCC, 2001).

The IPCC uses the concept of "radiative forcing" to compare the relative importance of different changes to the Earth's climate system. This is an effective average radiation (in units of $W/m^2$) entering or leaving the top of the troposphere (see chapter 1). Each radiative forcing requires the Earth to redo its energy budget such that there is an increase of temperature (a positive forcing) or a decrease (a negative forcing). The third report determined that the radiative forcing resulting from greenhouse gases was strongly positive. Even better, the IPCC report gave the forcing result with a relatively low uncertainty, $2.4 \pm 0.3$ $W/m^2$ and considered its level of scientific understanding to be "high." Considerable progress had been made with determining the effect of natural and manufactured aerosols. The IPCC found that the direct effects of all aerosols combined gave a somewhat negative radiative forcing. These effects ranged from negative 0.4 $W/m^2$ for sulfates to positive 0.2 $W/m^2$ for black carbon aerosols caused by fossil fuel burning. The uncertainties, however, were on the order of the effects themselves, and the IPCC still described its level of scientific understanding in this area to be "very low" (IPCC, 2001a, 8).

The third report also suggested that human influence would continue to alter the Earth's atmosphere, resulting in temperature increases throughout the 21st century. Although putting the global temperature rise of the 20th century at 0.6 degrees Celsius, the report put the possible rise in the next century as ranging from 1.4 to 5.8 degrees Celsius. It also suggested that $CO_2$ concentrations might go beyond a doubling of $CO_2$ by the end of the century. Along with these greenhouse gas and temperature increases would almost certainly come the continuing melt of the polar ice caps, the slow rise of average sea level (resulting from both the melting ice and the thermal expansion of water), and the possibility of species extinction and extreme weather.

With the third report of 2001, the hypothesis of anthropogenic global warming showed that it had weathered criticism during the 1990s from just about every quarter, ranging from those within the climate science community to the skeptics. The strengthening claims of the IPCC made waves in the media and so the issue gained greater public and political attention. Little of this new attention translated to significant action by the United States. Even though President Bill Clinton and Vice President Al Gore were sympathetic to the issues (Gore's environmentalist book, *Earth In the Balance* had come out in 1993), little changed for American policy regarding the environment and global warming. This is partly because conservative Republicans, many of whom strongly doubted either the fact or the importance of global warming, maintained control of Congress. And, as we saw in chapter 4, the U.S.'s disinterest and even hostility for the issue were especially noticeable during the administration of George W. Bush. Nevertheless, during the 1990s and early 2000s, the international community was moved to act.

## THE KYOTO PROTOCOL

As the scientific consensus deepened during the 1980s, other scientists besides James Hansen and John Houghton opted to enter the political area. Those who were convinced of the seriousness of the problem sought to pressure their governments to enter international agreements to reduce fossil fuel emissions. A series of international scientific conferences in Villach, Austria (organized by the United Nations Environment Programme, the World Meteorological Organization, and the International Council for Science) provided the opportunity. The report of the 1980 conference made relatively mild recommendations. Although it found that the potential threats of global warming called for an international program of cooperative research, there were still significant scientific uncertainties and no consensus on the severity of the problem. Therefore, attempting to specify $CO_2$ emission limits was premature.

Five years later, the Villach conference report, "Assessment of the Role of Carbon Dioxide and Other Greenhouse Gases in Climate Variations and Associated Impacts," went much further. It warned that greenhouse gases could be expected to cause warming in the next century that would be greater than any in human history. The report noted that "the rate and degree of future warming could be profoundly affected by government policies on energy conservation, use of fossil fuels, and the emission of greenhouse gases." It went on to recommend that government action seek a 20 percent cut in emissions by 2005. The 1985 Villach conference is widely credited with placing the issue of climate change firmly on the international political agenda. No doubt, it also added to pressure for the foundation of the IPCC, which followed three years later.

The next move toward an international agreement on limiting emissions came in 1992 at a special meeting of political leaders organized by the UN in Rio de Janeiro. Representatives from 155 nations attended this Earth Summit. Because climate change and environmental concerns had been brought to the attention of the political representatives attending, an agreement was signed by 150 of them. The UN Framework Convention on Climate Change (UNFCCC) did not ask for strict or significant targets for emission reductions, but it did represent a significant step forward. The participation of 150 of the world's nations suggested that they had the political will to revisit the issue.

Filling out the skeleton of the Framework Convention had to wait another five years, until the UNFCCC's third Conference of the Parties (COP-3) in Kyoto, Japan. This was a much larger affair than the one in Rio and included about 6,000 delegates from around the world, as well as representatives of a multitude of nongovernmental organizations from industry to the environment. Many governments, especially the Europeans, pushed for deeper and more aggressive action than had come out of Rio. The negotiations proved to be extremely difficult. The Europeans were relatively eager to reach an agreement, but other countries such as the United States were hesitant to support what they saw as restrictions on their economies. There also were difficult issues regarding comparisons of fairness in the requirements placed on the established industrial nations versus the developing nations. The passage of the Kyoto Protocol on December 11, 1997 was close. Last-minute negotiations were helped along by Vice President Albert Gore, who had a personal and political interest in Kyoto's success (Weart, 166).

The Kyoto Protocol has two fundamental aspects. First, it set carbon caps or budgets for each nation and second, it proposed a system of emissions trading designed to make the process responsive to free-market forces. Together, these provisions are often referred to as cap and trade. Let's consider each in turn. First, the targets for emission reduction varied according to the economic circumstances of each nation. The original targets for the most developed nations reduced their emissions to 5 percent of their 1990 levels by the period 2008–12. The target for the United States was slightly higher at 7 percent and that for the European Union higher still at 8 percent. Other countries were judged to be able to participate in the treaty, but because of economic limitations, would not be able to cut emissions. Russia was asked to stay even with 1990 levels, and Australia and Iceland were allowed an 8 percent and 10 percent increase, respectively. Because developing or poor nations were believed to be unfairly hindered by any such treaty, they were left out altogether and given no targets. In theory, their emissions could increase to virtually any level. As will be expanded on later, this raises concerns, as countries like China and India are positioned to be major polluters as their economies expand.

The protocol added the mechanism of emissions trading so that the actual emissions cuts might be relatively cost effective. The idea works simply. Consider trading within an individual nation. Once a particular company has exceeded the reduction target set by a particular national government, the government issues that company an emissions certificate. These certificates then act as a sort of "carbon money." Other companies that are having trouble meeting their targets can buy these certificates from the successful company. In this way, there is a strong economic incentive for meeting and exceeding the carbon targets. But there is also a stop-gap for companies that find fixing their emission problem hopelessly more expensive than buying the emissions permits from other companies (Houghton, 290). As of January, 2005, the European Union Emission Trading System enabled emissions trading among the various European nations.

After establishing the protocol in 1997, the next international milestone was reached at COP-6 of the UNFCCC, held in Bonn, Germany in July 2001. There, 186 countries agreed to sign and ratify the Kyoto Protocol. Unfortunately, just four months before, the new U.S. President George W. Bush announced that that the United States would not participate. This decision very well may have reflected the overall will of the American people at the time, but it seriously hampered the success of the treaty. The terms of the Kyoto Protocol were also somewhat weakened at the Bonn meeting. In the negotiations to convince countries like Japan, Australia, and Canada to sign, their targets were relaxed by a number of percentage points. Full ratification of the treaty had to wait until February 16, 2005, when Russia finally signed on, giving critical mass to the number of industrialized powers signing the agreement.

As might be expected, the Kyoto Protocol has met a great deal of criticism. In some ways, the most damning criticism is that it simply does not go far enough to have an effect on global warming. Whereas many scientists have estimated that cuts up to 60 or 70 percent will be necessary to keep $CO_2$ levels at a reasonable level (double that of preindustrial levels), the Kyoto Protocol asks much lower cuts of all nations participating in the treaty. Supporters of the treaty acknowledge this problem but see it as a first step. They hope that the Kyoto process will build up political will and public awareness of global warming and that the next treaty will be more ambitious.

In addition to Kyoto's lack of efficacy, there are doubts about its fairness. The different standards that were brokered for different nations sometimes appear to make little sense. The case of Australia, which compared to other nations is the worst per-capita emitter of greenhouse gases, seems especially egregious. The Australian representatives to Kyoto argued that their country had special limitations because of its unique infrastructure. More than most countries, Australia was forced to continue to rely on fossil fuels, high transportation costs (the continent has a relatively low population density), and an energy-intensive but profitable export market. To many onlookers, especially environmental groups, Australia brokered an unfair deal that exaggerated its emissions needs (Flannery, 222).

In addition, many critics noted the complete exclusion of the developing nations. Although the argument that the developed countries benefited in the past from cheap (and polluting) energy sources made sense, the fact remained that, in the present-day, developing nations such as China and India had a considerable advantage if they did not enter any agreement at all. Critics also pointed out that, if emissions reductions could really help the problem of global warming, then allowing major polluters to proceed without limit did not seem sensible. Indeed, in the 15 years since the 1992 conference in Rio, India has increased its emissions by about 100 percent and China by 150 percent.

Such facts seem to have been a roadblock standing in the way of U.S. participation. In turn, the nonparticipation of the United States has been a major problem for the treaty. Already in July 1997, the U.S. Senate unanimously passed the Byrd-Hagel Resolution, which stated that the United States must not sign any agreement that did not include targets for developing nations (Kolbert, 154). Knowing that the protocol had virtually no chance of passing the Senate, the Clinton Administration never submitted it for ratification. As mentioned previously, within months of taking office, George W. Bush announced that the United States would not sign. In spite of all this, local actions took root in the United States (see chapter 6). One of these was a voluntary cap-and-trade program called the Chicago Climate Exchange, which began in 2003. The more than 350 companies that joined the Climate Exchange agreed to reduce their emissions by 6 percent by the end of the first period (which ended in 2010).

A final set of concerns surrounds difficulties with the "carbon currency" established by emissions trading. At the moment, cap-and-trade systems are limited to national economies and the European Union. Many developing nations would like the carbon credit system to be extended to a fully international market as a means of improving their infrastructure and encouraging economic growth. With an international system of carbon trade, poorer countries could earn credits by updating their technology and agricultural practices, and engaging in reforestation. Such countries could then sell their carbon credits on the open market, thus generating further capital for development (Maslin, 130).

Other critics question the effectiveness of a cap-and-trade system. Many nongovernmental organizations and environmentalist groups question whether a market-based system can remain true to the moral ideals that inspired the response to global warming in the first place. And, indeed, there are many ways in which the cap-and-trade approach seems to miss the spirit of the law. For one thing, if the regulatory agencies issue credits too generously, then the market becomes flooded. In this case, although many companies can profit from trading the credits, polluters can too easily buy credits. This leads to negligible emissions reductions, which effectively eliminates the cap. Companies or countries might also try to push programs that achieve short-term improvements to earn carbon credits and therefore gain short-term profits. In so doing, they will likely bypass the more expensive options of scrapping old technology and investing in newer approaches. This sort of problem might actually stall the transition to less-polluting technologies. Finally, cap and trade appears to be open to manipulation, in which companies or countries pursue programs that will gain carbon credits but lead to unappealing results; one example is when old forests are cleared against the local population's wishes, in order to be replaced with new growth. Supporters of cap and trade tend to point out that any effort toward improvement of the emissions problem must somehow engage the profit motive.

Others suggest that additional policies might be used. One of these is a carbon tax, which would involve a national government or other authority levying a fee in proportion to the amount of $CO_2$ emissions. In this way the social cost of carbon (estimated as the cost to society of emitting one additional metric ton of carbon) would be paid for by the consumer by effectively raising the price of burning fossil fuel (Mann and Kump, 146, 156). Other observers, such as the political scientist David G. Victor, have argued for a hybrid approach that would combine aspects of emissions trading and taxation, thus allowing governments to make more flexible policy (Victor, 101).

## FURTHER SCIENTIFIC AND POLITICAL OBJECTIONS

From the time that scientists defined a consensus on anthropogenic global warming and entered the public arena in the late 1980s, their claims were inextricably tied up with political pressures and positions. This is not to say that they were in any way lying about their scientific results (although that possibility should not be discounted either). It is simply to say that any utterance on a scientific problem that potentially affects human society as much as global warming cannot avoid political entanglements. This Janus-faced nature of climate research, at one and the same time scientific and political, is as true of mainstream scientists supporting the hypothesis of anthropogenic global warming as it is of the many skeptics rallied against it. Both groups can be shown to demonstrate philosophical biases that go beyond the strictly scientific content of their work, and both have numerous ties to forms of funding and professional support that can always be used (fairly or unfairly) to call their judgments into question.

As chapter 2 reviewed, during the 1980s and 1990s, a number of scientific objections had been raised against the growing consensus, including inaccuracies in the global temperature record, doubts about the relative importance of solar variations, the possibility that increases in atmospheric $CO_2$ might actually be beneficial to humans, and inconsistencies between the different general circulation models. Some of these objections touched on areas of important ongoing research, whereas others represented little more than stubborn repetitions of questions that the climate science community had already addressed and answered to their satisfaction. And, as we have seen, the third IPCC report of 2001 gave strong conclusions compared to the first two. In little more than a decade of work, the IPCC had greatly strengthened its scientific arguments and had concluded that not only was global warming virtually an established fact, but that much of this was due to human activity.

This is not to say, however, that the hypothesis of anthropogenic global warming did not face further scientific challenges. Having a scientific consensus, or a shared paradigm, is not like having certain truth. As Thomas Kuhn has pointed out, every science paradigm contains unsolved problems and anomalies. To some degree, this is the normal state of ongoing research. Although the difficulties of a paradigm may sometimes become so great that it is eventually rejected (and another paradigm adopted instead), it is more common that anomalies motivate further work within the paradigm. So far, the latter has been true with regard to the consensus on global warming of the last quarter century.

One major anomaly arose regarding satellite determinations of global temperature. The satellite data, which first became available during the 1970s, must be considered a proxy measurement, as it begins with measurements of microwave radiation in various wavelength bands and then is back calculated (using known radiation laws) to obtain a value for temperature. When the first trends were analyzed in the late 1990s, the satellite proxy measurements showed a significantly smaller warming trend than ground-based measurements and possibly a small cooling. Temperature measurements from high-altitude balloons (available from 1958) more or less agreed with the satellite measurements and seemed to confirm the contradiction with ground-based measurements. This anomaly was noted in the IPCC's 2001 (third) report. During the next three years, a number of problems were identified with both the satellite and balloon data sets. The balloon readings turned out to contain a number of unreliable measurements because, over the years, there had been changes in balloon instrumentation and variations in how different groups analyzed data.

The satellite measurements faced a number of difficulties. First, problems were found regarding the calibration between the eight different satellites used to collect data. Second, the microwave measurements drifted slightly because the actual altitude of the satellites lowered slightly over the years (as a result of the slight friction between the satellite and the upper atmosphere). Third, the microwave measurements turned out to be sensitive to different altitudes, depending on the humidity of the upper atmosphere. Because the globe warmed over the last 20 years, the satellite measurements were effectively observing higher and higher altitudes. The Earth normally has a temperature profile with higher temperatures near the surface; therefore, this drift of effective measurement altitude introduced a spurious cooling trend (Mann and Kump, 38). By the time of the fourth IPCC report in 2007 (see the next section), the satellite and balloon measurements were consistent with ground-based measurements.

The debate over the satellite and balloon determinations of temperature has largely been settled, but other controversies have been more difficult. One of these concerns the possible connection between global warming and extreme weather. It is clear that the overall temperature trends of Earth's land and ocean surface temperatures have increased during the last three decades. In addition, reliable data show an overall increase in the severity of storms during past three decades. Both of these facts seem certain individually, but making a connection between them has been much more difficult. One of the first scientists to do so was atmospheric scientist Kerry Emanuel of MIT. During the summer of 2005 (shortly before Hurricane Katrina struck America's south coast), Emanuel published an article in *Nature* that strongly linked increasing storm strength to increasing ocean temperatures. A number of scientists challenged Emanuel's ambitious article. Scientists at the U.S. National Oceanic and Atmospheric Administration tended to believe that the increase in hurricane activity could be attributed to changes in ocean circulation that were unrelated to global warming. Meteorologist William Gray of Colorado State University made especially influential challenges based on his reputation as an expert on hurricanes. Gray's objections highlighted a difference of style between different climate scientists. Whereas Emanuel claimed a causal connection between global warming and increased hurricane strength based on extensive theoretical analysis and computer modeling, scientists like Gray tend to distrust modeling and to restrict their work to collecting and analyzing reliable data. They searched for correlations between data, but were hesitant to impute causation. As of the present writing, the causal connection between global warming and extreme weather claimed by researchers such as Emanuel has not been established unequivocally. Nevertheless, the consensus of the climate science community is that the connection between warming and more powerful storms is a serious enough possibility to warrant further scientific study and consideration when making public policy.

Consideration of the El Niño-Southern Oscillation (ENSO) further complicates the relation between global warming and future climate and weather patterns. For one thing, ENSO variability affects the global mean temperature anomalies. For example, the year 1998 was particularly warm. One reason for this appears to be that the winter of 1997–98 saw particularly strong El Niño events, which warmed a large part of the Pacific Ocean. Some skeptics have attempted to generalize this claim, saying that ENSO variations account for the lion's share of the increase of global temperature. Although variations of ENSO considered when interpreting global temperature trends, these variations have proven to be significantly smaller than the radiative forcing found for the increase of atmospheric greenhouse gases. Turning this around, there are also significant uncertainties regarding how global warming itself will affect ENSO. Most computer models indicate that ENSO is sensitive to warming trends, but the models do not yet agree on whether warming will lead to more El Niño-like or more La Niña-like conditions. In turn, uncertainty about ENSO increases uncertainty about extreme weather. William Gray has suggested that increased El-Niño activity leads (via conditions of higher vertical wind shear) to decreased Atlantic hurricane activity. Therefore, if global warming leads to more El Niño-like conditions, then this might bring fewer and not more Atlantic hurricanes (Mooney, 260).

Additional challenges to the hypothesis of anthropogenic global warming arose from more political channels. As discussed in chapter 4, a number of nongovernmental organizations and think tanks, such as the George C. Marshall Institute and the Global Climate Coalition, were founded during the 1990s. Newspapers with a conservative or business specialization tended to publish opinion pieces by the skeptics. The *Wall Street Journal,* in particular, has featured numerous editorials by S. Fred Singer, Frederick Seitz, Richard Lindzen, and its own editors.

In addition to such activities, the skeptics have generated a number of petitions and declarations. The Heidelberg Appeal of 1992 highlights a likely backdrop of these antiglobal warming efforts: a general opposition to the more radical elements of the environmental movement. The appeal claimed to be concerned about ecological problems, but it pleaded that these be addressed from a human-centered point of view, such that human needs would be balanced in a practical way against the ecological hazards created by human activity. The appeal denounced strident environmentalism as "an irrational ideology which is opposed to scientific and industrial progress, and impedes economic and social development." Although the wording of the appeal was not directed at global warming per se, this almost certainly was one of its targets, as it was released at the same time as the 1992 UN Earth Summit in Rio de Janeiro. It was eventually endorsed by about 4,000 scientists, including more than 70 Nobel Prize winners.

Other petitions were more specifically directed against global warming. The Leipzig Declaration was originally written in 1995 by S. Fred Singer, an atmospheric physicist who runs an organization called the Science and Environmental Policy Project. The short statement claimed that there was no scientific consensus regarding global warming. It also criticized political efforts like the 1992 United Nations Framework Convention on Climate Change for suggesting rash actions on the basis of incorrect science. The document claimed the signatures of about 80 scientists (including Frederick Seitz and Patrick Michaels). After the signers were contacted by a journalist, however, some claimed never to have signed the document. A significant number of others had no expertise in climate science, and still others had significant financial ties to the fossil fuel industry. Despite this setback, Singer updated the document in 1997 and then revised it in 2005; eventually it contained about 100 signatories (Science and Environmental Policy Project).

The Oregon Petition, the largest effort to deny a scientific consensus regarding anthropogenic global warming, was organized by the Oregon Institute of Science and Medicine. In 1999, the petition was circulated along with a cover letter written by Frederick Seitz. The petition itself was a brief statement saying that the Kyoto Protocol should be rejected and that the release of greenhouse gases by human activity could not be the cause of "catastrophic heating" (a term that the petitioners use as being more serious than mere "global warming"):

We urge the United States government to reject the global warming agreement that was written in Kyoto, Japan in December, 1997, and any other similar proposals. The proposed limits on greenhouse gases would harm the environment, hinder the advance of science and technology, and damage the health and welfare of mankind.

There is no convincing scientific evidence that human release of carbon dioxide, methane, or other greenhouse gasses is causing or will, in the foreseeable future, cause catastrophic heating of the Earth's atmosphere and disruption of the Earth's climate. Moreover, there is substantial scientific evidence that increases in atmospheric carbon dioxide produce many beneficial effects upon the natural plant and animal environments of the Earth. (Oregon Petition)

The petition and cover letter were accompanied by an article titled "Environmental Effects of Increased Atmospheric Carbon Dioxide" by Arthur B. Robinson, Noah E. Robinson, and Willie Soon. Although the article was revised in later years, the first version (dated 1997) challenged the claim of warming by using the early satellite data that was later shown to be incorrect. Somewhat more disturbingly, the article followed the same style and format as the scientific journal *Proceedings of the National Academy of Sciences*, even though the article was not published in a peer-reviewed journal. Whether it was the article's content or form, it apparently convinced quite a few people, and the petition Web site lists more than 31,000 signatories. Nevertheless, the content of the Robinson, Robinson, and Soon article has been largely rejected by the climate community. In addition, the usual problems exist with confirming the signatories of the petition and checking that their credentials pertain to climate science.

The seriousness of the global warming threat has been challenged by a number of full-length books but none so effectively and influentially as Bjørn Lomborg's book *The Skeptical Environmentalist,* published in Holland in 1998, and later in English translation (Lomborg, 2001). This was followed by a related book specifically addressing the topic of global warming (Lomborg, 2008). Lomborg's overall argument is directed not only at global warming but at a general type of overreaction to environmental problems that leads environmentally motivated scientists and politicians to propose impractical and wasteful policies. Although Lomborg accepts that global warming is occurring and that human beings have largely caused it, he wishes to show, in a cost-benefit analysis, that the cost of adopting measures like the Kyoto Protocol is much greater than the cost of addressing other more tractable problems (like AIDS, malnutrition, and malaria), which could benefit a greater number of people. He even suggests that global warming might bring some benefits to humankind, such as fewer people dying during the somewhat less cold winter months and increasing water supply as a result of the melting of the polar ice caps. Lomborg's work has been criticized on many grounds, including the fact that, as a political scientist with a specialization in cost-benefit policymaking, he has no expertise in climate science. Nevertheless, as a book on public policy, *The Skeptical Environmentalist* has had wide influence, particularly in the conservative press in the United States.

Despite ongoing scientific and political challenges during the 1990s and early 2000s, interest in the issue strengthened, even in the United States. As chapter 4 recounts, this came from a number of sources. First, a general increase of interest in environmental issues made the public more receptive. Second, at least part of the conservative Christian evangelical community showed itself to be interested in the idea of acting as the stewards of God's creation by taking action on global warming. Third, a number of businesses, especially in the insurance industry and those that might profit from mitigation and adaptation efforts, took account of global warming in their planning. In addition, the scientific consensus on anthropogenic global warming continued to deepen. Reviewing the fourth IPCC report of 2007 affords us a convenient way of surveying the state of the consensus (even though the IPCC is already at work on the next report).

## THE CONSENSUS RESTATED: THE FOURTH IPCC REPORT
### The Scientific Basis

As with the previous reports, the IPCC organized three working groups, one focusing on the scientific basis of global warming, another on the likely impacts of future climate change, and a third on the possible ways of mitigating these impacts. Each working group produced its own separate report, which we survey in the next three subsections. Much of the following discussion is based on the IPCC's Synthesis Report, although we have placed references to the three volumes (each totaling nearly 1,000 pages) produced by the three groups, at the end of each section (IPCC, 2007d).

The most certain aspect of the IPCC's findings is the changes in the Earth's climate, which it describes as "unequivocal." Of the previous 12 years (as of 2007), 11 of them were one of the 12 warmest years since the start of the systematic measurement of temperature (around 1850). Analyzing the 100-year trend in temperature, the IPCC found an average global increase of 0.74 degrees Celsius, which is about one-tenth of a degree more than the average found by the third IPCC report six years before. The temperature increases vary somewhat over the globe but are particularly strong in the highest northern latitudes. As a result, the extent of snow and ice in the Arctic regions has declined significantly, at a rate of nearly 3 percent per decade for the last three decades.

The temperature increases are backed up by a wide range of other indicators. For example, between 1961 and 1993, the IPCC found that sea level rose at about 2 millimeters per year. Since 1993, this number has increased another millimeter per year (although it is possible that this is only a temporary variation). Precipitation has also changed in ways that are consistent with the predictions of computer models; precipitation increased significantly in the eastern parts of North and South America, northern Europe, and northern and central Asia, and it has decreased in the Mediterranean region and the southern parts of Africa and Asia. Extreme weather also has increased, including heat waves, heavy precipitation, and intense tropical cyclone activity. Data on animal behavior also show systematic changes, especially regarding migration and mating behavior, which are now occurring earlier in the spring. In sum, nearly 90 percent of the data sets regarding changes in physical systems (involving snow, ice, or water) and biological systems (including terrestrial, marine, and freshwater life) show trends that are consistent with warming.

Of course, it is easier to be certain about basic climate trends than it is to identify the causes of climate change. Nevertheless, the fourth IPCC report offers significantly stronger conclusions on causation than the three earlier reports. It is virtually certain that the concentrations of greenhouse gases (GHGs) in the atmosphere have increased well beyond preindustrial levels, from about 280 ppm during the 18th century to about 380 ppm today. It is also virtually certain that the release of GHGs (including $CO_2$, methane, nitrous oxide, and CFCs) caused by human activity has increased during the last century (by about 80% between 1970 and 2004). This release occurred through the direct burning of fossil fuels but also as the result of land clearing. Most important, the IPCC report says that it has "very high confidence" that these two facts are connected: the added GHG levels observed in the atmosphere can be traced back to the release of GHG by human activity.

The next question that must be addressed is whether the GHGs released to the atmosphere by human activity can explain the temperature data and other trends associated with global warming. This is the point at which the climate community and the IPCC must rely on computer models to some extent. Although it is possible to do hand calculations to show that an increase in GHGs can produce global warming (such as those done by Svante Arrhenius 100 years ago), one must resort to numerical modeling to get a reliable picture. General circulation models contain most of the major climate processes (including, the global radiation budget, atmospheric circulation, and oceanic circulation, clouds, and aerosols, and changes in land use). After comparing the results of the best computer models available, the IPCC found that two of the most important natural causes of climate change, solar variations and volcanoes, would have produced a small cooling over the past few decades. When anthropogenic factors were added (mainly the release of GHGs and changes in land use), however, the model results were consistent with the observed global warming trends.

Stated in more technical terms, the negative radiative forcings were relatively small compared to positive radiative forcings. To compare some of the more important of these: The effect of added $CO_2$ is estimated to be equivalent to a radiative forcing of $+1.66 \pm 0.17$ W/m$^2$, whereas that caused by other GHGs (including $CH_4$, $N_2O$, CFCs, and ozone) is about $+0.94 \pm 0.30$ W/m$^2$. Other factors are less well known. For example, the net effect of changes of the surface albedo (caused by changes in land use and snow cover) is about $-0.1 \pm 0.2$ W/m$^2$. Finally, the overall effect of aerosols is about $-1.2 \pm 1.0$ W/m$^2$. Note that in the latter two cases, the uncertainty is on the same order as the effect. When adding together all known anthropogenic causes, the net radiative forcing is about $+1.6 \pm 0.9$ W/m$^2$. By contrast, the radiative forcing caused by the most important natural factor, variations in the total solar irradiance (the amount of solar radiation reaching the top of the Earth's atmosphere), although positive, is much too small to account for global warming trends, $+0.12 \pm 0.06$ W/m$^2$. This estimate is less than half that given in the third IPCC report.

In addition to average global trends, computer model results used by the IPCC do a good job reproducing regional climate changes over the last 100 years. Model results were compared to temperature data for the continents of North America, South America, Europe, Africa, Asia, and Australia. In each case, the use of natural forcings only in the computer models undercut the actual temperature record. But when anthropogenic forcings were included, the model results agreed well with the data, within uncertainties. Beyond reproducing global and regional temperature records, the model results were consistent with other data regarding sea level rise, changes in global wind patterns, and the increase of extreme weather, including extremes of temperature, droughts, and storm (IPCC, 2007a).

## Impacts

As we have pointed out, the computer simulations of climate change cannot claim to be "predictions" of the future. For this reason, the IPCC uses the terms *forecasts* and *scenarios* to describe the model results. Beyond the uncertainties regarding the climate data and physical processes, there is considerable and unavoidable uncertainty about future social and economic conditions. Social and economic conditions are of crucial importance, for they have a major effect on future emissions and also the ability of humans to mitigate the effects of warming. The IPCC assumes different sets of assumptions about the future state of human society, which are then translated to the raw data given to the computer models. Each scenario is designated by a code. There are 35 in all, but each of them is based on one of four basic "storylines": A1, B1, A2, and B2. The A1 storyline proposes a future in which the world's nations are relatively integrated and able to make treaties and other agreements. At the same time, global capitalism will continue to expand, as new and more efficient technologies enable rapid economic growth. This is expected to result in a reduction in regional differences in per capita income. It is also assumed that the world's population will grow until about the mid 21st century, at which point it will slowly decline. The B1 storyline also proposes a relatively integrated world, with a population that peaks at mid century. The difference is that this storyline is a future in which the economy shifts, moving toward a service and information economy. This, in turn, will allow the adoption of clean and resource-efficient technologies, as well as a policy emphasis on environmental sustainability.

By contrast with either of the first two storylines, the A2 storyline offers a divided world, in which nations and regions seek to be relatively independent, both politically and economically. In this case, the world population is expected to grow steadily but more slowly than the other storylines. In addition, economic and technical growth is expected to be relatively slow. Like the A1 storyline, ecological issues will not be at the forefront of public policy. The B2 storyline, like A2, concerns local solutions but, like B1, assumes a world that is concerned about environmental sustainability. This scenario is expected to involve relatively slow population growth and economic development (Houghton, 140).

Although the projections for the period 2000 to 2100 show warming for all of these scenarios, they give a wide range of results, centered around 1.8 to 3.6 degrees Celsius. As one might expect, the temperature projections for the ecologically friendly "B" futures tend to be somewhat lower than the "A" forecasts. Also, the international cooperation of the "1" futures tend to yield lower temperature forecasts than the "2" scenarios.

The projected rise in sea level by 2100 also varies greatly, from about 0.2 to 0.5 m. These changes will be due to the thermal expansion of water and the melting of the ice caps. This second factor gives especially large uncertainty to the estimate of sea level rise. Although Antarctica has lost some of its coastal ice (most notably the Larsen B ice shelf, which disintegrated in early 2002), it is likely to remain much colder than the Arctic regions, which are of greater concern. Greenland has already seen significant ice loss and will continue to lose during the 21st century. Because Greenland's ice is on land (built up as the result of centuries of snow fall) and not floating on water, its complete melt would raise the world's sea level by some seven meters. Although there is considerable uncertainty about how fast this might happen, current estimates suggest it might take several centuries.

The computer models also give projections about how different types of global resources will be affected over the next 100 years. Although the magnitude of the change on each sector varies strongly with the scenario, the general trends are clear. The availability of fresh water is changed by warming trends in ways that benefit some areas and hurt others. Availability will increase in the tropics and the high latitudes, but it will decrease in the mid latitudes and the semiarid areas of the low latitudes. Most of the changes to ecosystems will be damaging, as plants and animals must respond to changes faster than evolution allows. Even at a level of warming of 2 degrees Celsius, it is projected that 30 percent of the world's species will be extinct. Scenarios in which warming is greater and exceeds 3.5 degrees Celsius result in higher extinction rates, some reaching 70 percent.

As the oceans warm up, they will continue to act as a sink for $CO_2$ in the atmosphere. Therefore, marine life will also be affected. Most noticeably, coral reefs will be damaged as higher water temperatures and higher acidity (brought on by higher $CO_2$ concentrations) lead to bleaching. Corals reefs are formed by large colonies of tiny polyps, as they secret calcium carbonate to build their hard coral skeleton. Corals obtain most of their nutrition by virtue of a symbiotic relationship with unicellular algae that undergo photosynthesis (this is one reason that corals grow in clear, shallow water). The warming and acidification of the oceans cases the death or expulsion of the algae, which shows up as a whitening (or bleaching) of the coral reefs.

The loss of some of our terrestrial, freshwater, and marine species may or may not seem important when seen from a conservationist perspective. Some people may not inherently value the beauty and diversity of the Earth's flora and fauna. One must also consider, however, that the balance of the ecological system is important to its ability to provide human beings with crucial goods and services. Food is of special concern. Because of the increasing aridity at lower latitudes, the productivity of growing cereals (rice, maize, wheat, oats, and so on) will decrease, although it may increase to a degree at higher latitudes. This type of change may or may not result in a net reduction of food production, but it certainly stands to disrupt the regions where this production presently occurs, and therefore will disrupt human societies and economies. It is also projected that warming will disrupt relatively small producers of food such as subsistence farmers and fishers.

The general trend in precipitation is expected to be an increase, because global warming will increase evaporation. But changes to precipitation will vary with region, as the major circulation cells of the atmosphere are expected to change; the Hadley, Ferrel, and Polar cells, along with the jet streams, will likely shift toward the poles of both hemispheres. This will result in a general pattern of precipitation changes: an increase at the equator, a decrease in certain areas of the subtropics (the region of the Hadley cell), a decrease in certain areas of the middle latitudes (the region of the Ferrel cell), and an increase in the polar regions. In North America, the lower latitudes are expected to experience lower precipitation and a higher incidence of drought. Some of the higher latitudes are expected to see increased rainfall, which might benefit select agricultural crops. Other crops that are already at the warm end of their suitable temperature range, however, will be weakened.

The link between global warming and extreme weather remains uncertain. Nevertheless, the IPCC considers it "more likely than not" that the two are linked and that it is "likely" that extreme weather will increase during the 21st century. North American cities will probably experience more frequent and severe heat waves and coastal communities will be subject to greater flooding and storm activity. Europe will be somewhat more adversely affected and experience greater differences between its regions in terms of seasonal variation, amount of precipitation, and extreme weather. The southern countries will be subject to higher temperatures and reduced water availability and, therefore, worsening crop productivity. The many mountainous areas will see reduced snow cover and glacier retreat, resulting in significant species extinctions and loss to the tourist industry. Once again, coastal areas will be subject to more extensive and frequent flooding. The melting of the glaciers and ice in the Arctic will certainly require significant changes in infrastructure (as roads and buildings sink as a result of melting permafrost) and likely lead to a loss of many traditional, indigenous ways of life.

Parts of Africa and Asia will be hard hit by climate change, in part because their relatively poor economies and large populations might have difficulty in responding quickly. Both continents feature numerous low-lying coastal regions that are highly populated and will be affected by rising sea levels. At the same time, the availability of fresh and potable water may decline significantly. In Africa, there may be an increase of arid and semiarid land (by more than 5%). Reductions in rainfall might reduce yields from rain-fed agriculture by as much as 50 percent. This, in turn, will put increased pressure on the African populations, who are likely to suffer from greater malnutrition. Asia will similarly be subject to increased flooding in some populated regions but increased drought in others, and an increased incidence of malnutrition and diarrheal disease (IPCC, 2007b).

## Mitigation

The third working group concentrated on possible ways of mitigating future warming trends. In deciding which mitigation efforts to pursue, it is important to minimize cost-benefit ratios and to account for the fact that certain nations face poverty and unequal access to resources and food. On the one hand, some of the mitigation approaches may raise objections along the lines of Bjørn Lomborg's, that they will be too expensive and that the resources would be more efficiently spent on other problems. On the other hand, it is likely that many initiatives will be "win-win" propositions, in that they not only halt or slow the increase of greenhouse gases but also address other environmental problems and offer new sources of wealth.

Both the second and third working groups address the sometimes complementary and sometimes contradictory relationship between mitigation (the effort to stop or slow the rate and magnitude of global warming) and adaptation (the effort to lessen the impact of climate changes on human society). One example of a contradiction is when a new adaptation technology requires the use of large amounts of energy, thus complicating mitigation. Nevertheless, certain adaptation efforts will be almost impossible to avoid, such as those concerning water supplies. Nations with flooded coastal areas will have to consider particularly expensive solutions, from building or rebuilding seawalls and storm surge barriers (as has already happened in the Netherlands) to population relocation (as has happened to a number of Pacific Islands). Agricultural practices will also have to be significantly modified and perhaps crops entirely relocated as the result of new global patterns of precipitation, erosion, drought, and earlier spring seasons. As warming continues and the global population also increases (it may exceed 9 billion by the year 2050), there is likely to be water shortages, especially in the developing nations. It will therefore be necessary to expand efforts to harvest rainwater, to desalinate ocean water, and to find new ways of reusing waste water.

Many technological improvements have been proposed as part of the effort to mitigate future global warming. Some of the most important involve the generation of energy. Of all energy sources, coal is one of the least desirable, as it releases more $CO_2$ per unit of generated energy than even oil. For this reason, the days of coal burning on a large scale are probably numbered. Nevertheless, so-called clean coal technology hopes to extend the viability of coal by limiting the release of $CO_2$ from coal-burning plants. Before burning, a portion of the minerals and impurities residing in the fuel are chemically removed. After the fuel is burned, the escaping flue gases are "scrubbed" with further chemical processes, which remove a number of pollutants, one of which is (a portion of) the $CO_2$ content. Although such technology can reduce emissions, it increases the cost of coal burning and also creates its own waste products. In carbon sequestration schemes (see chapter 6), the recovered $CO_2$ must be stored in a repository (often underground).

Other fuels are more promising than coal or oil, as they produce less $CO_2$ in the first place. Natural gas (which is primarily methane) has become the preferred fuel in the power industry worldwide. It burns hotter and cleaner than coal or oil but has the drawback of being harder to store because of its low density (it's a gas!). In addition, it offers only limited improvements; if all power generation were done by natural gas instead of coal it would only represent an improvement in $CO_2$ emissions of about 30 percent. There are hopes that hydrogen gas might be a step beyond natural gas, as it burns even more cleanly than methane. Unfortunately, there are daunting technical problems to surmount. For one thing, hydrogen requires more complex devices for burning than natural gas, including so-called molten carbonate fuel cells and proton-exchange membrane fuel cells. Storage is also a difficulty. A hydrogen fuel tank for a typical automobile would require pressures on the order of 5,000 psi (about 100 atmospheres). There are also concerns that hydrogen gas will be relatively expensive to produce on a mass scale.

A number of other power sources are being considered. Nuclear fission power, using uranium as a fuel, is one hope. The associated plant technology has improved tremendously since the time of the Three Mile Island and Chernobyl accidents, in 1979 and 1986, respectively. Unfortunately, the problem of what to do with the dangerous radioactive waste products produced by the fission reactions is still largely unsolved. Waste products must be buried or otherwise contained, although the amount of waste can be reduced somewhat with reprocessing technology. Nuclear fusion machines have been investigated during the last 50 years (using tritium and deuterium as fuels). Although such machines would generate less dangerous waste, none have been successful as a power generator; as of this writing, they require more power to run than is generated from fusion reactions.

Some of the most appealing alternative power sources are renewable sources, so-called because their rate of production by natural processes is similar to their rate of consumption by human beings. The generation of power using wind turbines has increased greatly in the last decade but has a few technical limitations that will keep the percentage of power delivered by the wind to a relatively small number (worldwide, it is currently 1.5%). For one thing, wind power is intermittent, as the turbines spin only when the wind is blowing. In addition, wind power is nondispatchable, meaning that to be economically viable, the output from the turbine must be used at the same time that it is generated. Solar power is a second promising renewable energy source. The most common form of solar power is the use of semiconductor-based photovoltaic cells. These have been used successfully in special, small-scale applications such as satellites, but are harder to use in larger applications. The use of solar power technology will probably increase significantly in the future, but at present it supplies only about a 50th of a percent of the world's energy. Other renewable sources include geothermal (taking advantage of heat flow from the interior of the Earth), bioenergy (which can be derived from sugarcane and agricultural waste products), and tidal and wave energy (using underwater turbines that generate power in somewhat the same way as water passing over a waterwheel).

Further mitigation technologies can be found in transportation, where it is generally desirable to shift away from gasoline-based internal-combustion engines. Diesel engines are more efficient and enable the combustion of a wider range of fuels, such as biodiesel fuel, which is derived from vegetable and animal fat. Automobiles based on hybrid engines (the most common right now is a combination of an internal combustion engine and an electric motor) get roughly double the gas mileage compared to gasoline-based automobiles. A further option is to shift wherever possible to rail and other forms of public transportation. Still more savings are possible whenever people are willing to shift to nonmotorized transport such as bicycles and walking.

Many improvements can be made to infrastructure, and these vary in practicality and cost effectiveness. Building codes can be defined such that new construction makes use of more efficient electrical appliances and heating/cooling systems. Factories also might be required to use systems that recover waste heat and power and recycle otherwise wasted manufacturing materials. Improvements can be made in agriculture by adopting new methods of crop and grazing land management such that carbon is better stored in the land and not re-released to the atmosphere. The release of methane can also be limited by improving rice cultivation techniques and livestock and manure management. In forestry, efforts can be made to avoid deforestation, to encourage reforestation, and to use dead vegetation and grasses to produce biofuels.

In closing, we highlight something noted by the IPCC and many other observers: Any successful effort to mitigate and adapt to global climate change will certainly not have a single answer. Instead, a realistic effort will probably involve a (possibly changing) menu of responses, many of which might be drawn from the ones mentioned in this subsection. The trick will be in finding a collection of initiatives and policies that will tackle the problem, despite the fact that each element of the response, when seen individually, might appear inadequate (IPCC, 2007c; Houghton, 325).

# 6

# Present and Future Policy Options

With the passage of the Kyoto Protocol by many nations, efforts began around the world to craft policies to create initiatives at the local, state (or regional), or federal levels. The United States was very much a different case. The issue of global warming engaged some of the touchstone social and cultural topics that defined constituents' political positions. Inserted into an already charged era of "culture wars" (in which environmentalism was associated with the radicalism of the 1960s), the debate over climate change was mired for more than a decade on establishing whether or not the phenomenon existed at all, let alone whether or not humans bore a significant responsibility for causing it. Earlier chapters have demonstrated the growing scientific consensus that gave way to international policy initiatives. Although American policymakers found it difficult to construct a base of support for climate legislation, much of the public was ready to act. With a cultural foundation shaped by modern environmentalism, a growing number of Americans moved forward to the even thornier issue: What should be done about climate change?

Because the federal government began the 21st century denying the global predicament that was bringing many other nations to action, reform activity in the United States primarily began at the local levels with communities and individuals who expressed their displeasure with government inaction and then demanded a response. Hundreds of localities defied the reaction of the federal government and made good-faith efforts to meet the conditions and expectations of the protocol, even though the American towns were not bound to Kyoto in any formal way. By 2007, large-scale initiatives were being built on this grassroots framework. States and regions worked to craft and adopt climate policies. Internet sources soon began to create a network of good ideas for communities to put into practice, including the development of regional greenhouse gas reduction markets, the creation of state and local climate action and adaptation plans, and increasing renewable energy generation. The major appeal for these changes was to reduce emissions that contribute to climate change; however, policymakers also found supporters of initiatives to reduce the nation's vulnerability to volatile energy prices, as well as to stimulate state and regional economic development and to improve local air quality. As national efforts take shape, these state and regional climate initiatives may provide models. They may also provide an existing infrastructure that some experts speculate might aid the United States in catching up with nations already moving in a carbon-neutral direction.

Despite the fact that the federal government has been slow to respond to the challenge of global warming, there are signs that it is now considering the issue in making policy. The *New York Times* headline for an August 8, 2009, story by John M. Broder called attention to a leading threat to national security, but another nation was not the source of these potential problems: "Climate Change Seen as Threat to U.S. Security." Referring to a newly released State Department report, the article opens:

The changing global climate will pose profound strategic challenges to the United States in coming decades, raising the prospect of military intervention to deal with the effects of violent storms, drought, mass migration and pandemics, military and intelligence analysts say. Such climate-induced crises could topple governments, feed terrorist movements or destabilize entire regions, say the analysts, . . . who for the first time are taking a serious look at the national security implications of climate change. Recent war games and intelligence studies conclude that over the next 20 to 30 years, vulnerable regions, particularly sub-Saharan Africa, the Middle East and South and Southeast Asia, will face the prospect of food shortages, water crises and catastrophic flooding driven by climate change.

The new seriousness with which this issue has been perceived has trickled into the expectations of the American public and, therefore, into the official efforts pursued by states and localities throughout the United States. The rest of this book offers one interpretation of the origin of these changes. This chapter focuses on a variety of U.S. initiatives at the regional, state, and local levels. It also describes some of the organizations that are focused on getting information on such initiatives to interested political leaders, as well as some of the leading business initiatives that such legislation is stimulating. Because the United States is playing catch-up with many of these initiatives, however, this chapter must also enumerate a selection of those implemented at various levels around the world. Finally, we survey a few of the policies and trends of the emerging economy of climate change.

## REGIONAL AND STATE INITIATIVES

In his 2004 book *Statehouse and Greenhouse,* political scientist Barry Rabe wrote that in the previous decade the American federal government had accomplished "virtually nothing" on climate change. This reality, though, provided context for his main point as he continued:

Alongside the extremist rhetoric and the barriers to serious engagement of this issue at the federal level, an almost stealth-like process of policy development has been evolving. . . . over the past decade approximately one-third of the American states have enacted multiple policies that show considerable promise of reducing greenhouse gases. . . . they demonstrate that it is politically possible in the United States both to form coalitions to support initiatives to reduce greenhouse gases and to take initial steps to secure implementation. (Rabe, xi–xii)

In short, while media pundits and national politicians debated the issue of global warming, a growing number of state leaders got to work. Among these initiatives, it is not surprising that a leader has been California.

After being a leader on so many earlier environmental issues, California in 2005 created a Climate Action Team (CAT) that would advise the governor on how the issue should impact planning in every agency. The CAT's primary responsibility is to coordinate the state's emission reduction programs and to write periodic reports on how well the state has met the greenhouse gas targets established by the Global Warming Solutions Act of 2006. This Act seeks to reduce California's emissions to 1990 levels by the year 2020. In pursuit of this sizable shift in energy use, the CAT has stressed across-the-board participation by each segment of state government. For instance:

The California Integrated Waste Management Board (CIWMB) addresses climate change issues through recycling programs, which avoid emissions from the energy-intensive processing and, instead, substitute sustainable building activities that emphasize energy, water, and materials efficiency thereby reducing emissions from their production and transport; and through landfill gas collection, which directly uses landfill greenhouse gas emissions for fuel. (CA Climate)

In addition, each state park is assessing its use of resources in order to reduce its "carbon footprint." To stir localities to action, the Public Policy Institute of California (PPIC) releases a study each November that examines how cities and counties are responding to climate change. In assessing climate policy at the local level recently, the PPIC found that an astonishing 75 percent of California's cities and counties are actively engaged in climate change issues.

"From wind energy in Texas to carbon dioxide standards in New Hampshire and Oregon," writes Rabe, "the elements of a bottom-up American approach to climate change are beginning to take shape" (Rabe, 109). Other states such as New York, Oregon, and West Virginia have passed lists of goals for limiting and adjusting energy production. Most of these regulations require growing percentages of energy to be acquired from renewable sources. In each state, however, politicians must consider the existing economic structure of their states. For instance, a state such as Oregon can develop standards for greenhouse gasses that oversee the full lifecycle of transportation fuel, including production, storage, transportation, and combustion and any changes in land use associated with the fuel. In addition other programs in Oregon reduce emissions from transportation by including requirements to reduce aerodynamic drag in medium- and heavy-duty vehicles, tire efficiency standards, and idling reduction requirements for trucks and ships.

In the coal state of West Virginia, for instance, the 2009 change sets a goal for the state to acquire 25 percent of all energy from renewable sources by 2025. In addition to traditional alternatives such as solar, geothermal, and wind, the West Virginia legislation allowed for the use of "advanced coal technology (e.g., carbon capture and storage, supercritical technology, ultrasupercritical technology and pressurized fluidized bed technology), coal bed methane, natural gas, fuel produced by a coal gasification or liquefaction facility, synthetic gas, integrated gasification combined cycle technologies, waste coal, tire-derived fuel, pumped storage hydroelectric projects, and recycled energy." The bill sets up a system of tradable credits—similar to "cap and trade"—to implement the program, and required utilities to hold credits equal to their generation. One credit is awarded for each megawatt-hour of alternative energy generation, two credits for renewable energy generation, and three credits for renewable energy generation located on a reclaimed surface mine.

Particularly since Rabe's book was published, the unique details of climate change have led to an innovative series of efforts on the environmental policy front. In particular, new efforts in transborder cooperation have defined this new era. As the Pew Center for Climate Change points out:

States and regions across the country are adopting climate policies. These actions include the development of regional greenhouse gas reduction markets, creation of state and local climate action and adaptation plans, and increasing renewable energy generation. In addition to addressing climate change, states and regions pursue these policies to reduce their vulnerability to energy price spikes, promote state economic development, and improve local air quality. State and regional climate policy will provide models for future national efforts, achieve greenhouse gas emissions reductions, and prepare for the impacts of climate change. (Pew, About U.S. States and Regions)

Regional efforts have linked states together in every area except the American South. The Pew Center goes on to describe a number of these initiatives:

On February 26, 2007, Governors Napolitano of Arizona, Schwarzenegger of California, Richardson of New Mexico, Kulongoski of Oregon, and Gregoire of Washington signed an agreement establishing the Western Climate Initiative (WCI), a joint effort to reduce greenhouse gas emissions and address climate change. Since that time, the governors of Utah and Montana, as well as the premiers of British Columbia, Manitoba, Ontario, and Quebec have joined. . . .

On November 15, 2007, six states and one Canadian province established the Midwest Greenhouse Gas Reduction Accord (MGGRA). Under the Accord, members agree to establish regional greenhouse gas reduction targets . . . and develop a multi-sector cap-and-trade system to help meet the targets. Participants will also establish a greenhouse gas emissions reductions tracking system and implement other policies . . . to aid in reducing emissions. . . . The Governors of Illinois, Iowa, Kansas, Michigan, Minnesota, and Wisconsin, as well as the Premier of the Canadian Province of Manitoba, signed the Accord as full participants; the Governors of Indiana, Ohio, and South Dakota joined the agreement as observers to participate in the development of the cap and trade system. . . .

On December 20, 2005, the governors of seven Northeastern states announced the creation of the Regional Greenhouse Gas Initiative (RGGI). . . . The governors of Connecticut, Delaware, Maine, New Hampshire, New Jersey, New York, and Vermont signed a Memorandum of Understanding agreeing to implement the first mandatory U.S. cap-and-trade program for carbon dioxide . . . The program will begin by capping emissions at current levels in 2009, and then reducing emissions 10% by 2019. Pennsylvania and the District of Columbia are observers in the RGGI process. On January 18, 2007, Massachusetts Governor Deval Patrick signed a Memorandum of Understanding committing his state to join RGGI. . . . In his State of the State address on January 30, Governor Donald Carcieri announced that Rhode Island would also be joining RGGI. On April 6, 2006, Maryland Governor Robert L. Ehrlich Jr. signed into law the Healthy Air Act. The bill required the Governor to include the state in RGGI by June 30, 2007. Maryland became the 10th official participating state in April 2007 with Governor Martin O'Malley's signing of the RGGI Memorandum of Understanding. (Pew)

As noted in the next section, many of these state and regional efforts were often inspired and modeled on international examples. In addition, some

U.S. communities were moving ahead at the local level with the help of a variety of initiatives and nongovernmental organizations.

## PUTTING LOCAL EFFORTS TOGETHER

In hundreds of cities, suburbs, and rural communities across the United States, frustration with a lack of federal action over the last decade has turned to new initiatives on climate change. We have seen the activities of this network at the state and regional level; however, the true grassroots action is at the local level. In 2002, Burlington, Vermont started a program to bring about a 10 percent reduction of greenhouse gas emissions. The town encouraged the use of renewable energy, recycling programs, and a variety of energy-saving practices (Kolbert, 171). In 2006, Boulder. Colorado passed a referendum to levy the nation's first climate tax in which electricity users are charged an extra fee based on how much energy they use. The proceeds are used to support the Boulder Climate Action Plan, which coordinates the city's emissions reductions efforts. Many other cities pursued small but significant steps. Fargo, North Dakota replaced all of its traffic light bulbs with light-emitting diodes (which use 80% less energy). Carmel, Idaho replaced its city fleet with hybrids and vehicles that run on biofuels. And Chicago has encouraged the planting of rooftop gardens, which reduce the need for air conditioning (Simon, 2007).

The Environment page of the citymayors Web site credits Seattle Mayor Greg Nickels with calling on mayors across the country in 2004 to take action that would coincide with the February 2005 date when the Kyoto Treaty came into legal force in nations that signed it. This attracted an immediate response and led to the adoption by the U.S. Conference of Mayors in June 2005 of a climate protection agreement calling for a 7 percent reduction by 2012 from 1990 emissions levels. The idea has spread through many cities, as noted by the citymayors Web site, which reads: "By [2005], there were already 161 signatories. Since 2007 this agreement has been supported by the Mayors Climate Protection Center and by March 2009 there were 935 signatories from cities containing approximately 84 million people" (http://www.city mayors.com/environment/green-mayors.html).

Creative thinking is partly responsible for these initiatives; however, a variety of organizations have prioritized finding ways to deliver new ideas and policy descriptions to local leaders who may have begun knowing little about climate change. For instance, Cornell University's Ag Extension Program provides a planning service to surrounding communities on how to implement policies at the local level that will address climate change. Among the recommended strategies for Tompkins County, N.Y., a Cornell report lists: "better manage forests, including harvesting wood and other 'green' biomass to supplement existing coal, oil and gas supplies; synchronizing traffic signals; creating incentives for carpooling; turning home thermostats down to 65 degrees (which would save the average resident an estimated $1,400), running city vehicles on gas-ethanol mixtures; and using wind power to generate some of the county's electricity needs" (Chambliss).

On the national level, the Pew Foundation may be the leader in educating the public at a variety of levels. Its Web site, Climate Techbook (http://www.pewclimate.org), acts as a clearinghouse for "low-carbon technologies—climate solutions at your fingertips." This site provides information about climate change that can be used by local officials to demonstrate the need for action to their constituents. It also illuminates efforts undertaken by communities throughout the world. On its page "Communicating Climate Change," the Pew Foundation explains:

Policymakers, businesses, the media, and the public are increasingly interested in this complex issue. This generates a multitude of information and opinions about climate change, and presents significant challenges to accurately and effectively communicate the issue. (Pew Foundation)

In other portions of the site, the foundation runs a type of policy "scorecard" to help keep interested parties abreast of the status of various initiatives.

Other organizations have specifically tried to make connections between localities throughout the world, breaking down the borders of nations in order to attack a shared problem. Although these organizations place great significance on international and national policy initiatives, many reflect one of the original mottos of environmentalism: "Think Globally, Act Locally."

One last model is that of Kyoto USA, an organization whose Web site functions as a clearing house for initiatives being tried by cities and towns anywhere in the world. The site seems to play on the competitive nature of political leaders by maintaining a running tally of which cities have committed to Kyoto and what initiatives they have put into practice. Because many other communities worldwide have instituted policies related to climate change, Kyoto USA uses the information on its Web site to alert U.S. mayors to efforts all over the world. It describes itself as: "an all volunteer, grassroots organization that encourages U.S. cities and their residents to reduce the global warming greenhouse gas emissions for which they are responsible" (http://www.kyotousa.org/). In its "Tools for Climate Activists," the site offers more actual examples of initiatives underway.

## GLOBAL MODELS LEAP AHEAD

Undeniably, U.S. communities are playing catch-up with efforts going on elsewhere in the world. One of the reasons for this, however, is not the domestic political activity in the United States; instead, a portion of the blame can be attributed to the overt political efforts of global organizations, particularly those brokered through the United Nations (UN). The UN, through the initiatives of the Intergovernmental Panel on Climate Change (IPCC) and sustainable development agencies, extended the global reaction well beyond that possible in the United States. and other developed nations.

Many of these initiatives grew from the fourth IPCC report of 2007 (see also chapter 5), which stressed that the responses of society to climate change will require aspects of both mitigation (slowing and lessening global warming itself) and adaptation (lessening the impact of climate change on society). The IPCC seeks to identify policies that lead as much as possible to "win-win" relationships among mitigation, adaptation, sustainable development, public health, and economic growth (despite the frequent and unavoidable tradeoffs). Some of these policies, in order to function effectively, must be imposed by a central authority at a national or international level, including government regulation, tradable permits (cap and trade), and price signals (additional taxes). Other policies are somewhat less top-down, including incentives (tax breaks and grants) offered to companies or nongovernmental organizations and the funding of research, development, and demonstration efforts. Still other initiatives are voluntary and originate at the local and regional level. Nevertheless, these often can be encouraged at the national and international level. For example, governments can pursue "information instruments" that simply disseminate information regarding best practices to firms, consumers, and nongovernmental organizations (IPCC, 2007c, 745).

Community action and urban planning have become some of the most dramatic efforts to create policies to help combat climate change. UN Secretary-General Ban Ki-moon has maintained that the issue of climate change is not just an environmental issue, but one that has serious social and economic implications as well. UN initiatives, therefore, have stressed a variety of sectors, including finance, energy, transport, agriculture, and health. UN policymakers have sought a way of creating overarching policies that stimulate and continue existing local efforts. In the optimistic atmosphere before the fifteenth Council of the Parties (COP-15) in Copenhagen, for example, the UNFCCC Negotiators released a memorandum titled "Climate Change Adaptation Strategies for Local Impact." After expressing hope for the adoption of a substantial agreement at Copenhagen (see the Epilogue for an update), the document stressed the importance of local action:

Yet ultimately success in adaptation must be measured in terms of impact on the ground at local level. Compared to climate change mitigation, climate change adaptation policy development is still in its infancy. Since adaptation was put on an equal footing with mitigation in the Bali Action Plan, significant progress in policy development can be observed. Yet the main focus of the debate is on the development of national adaptation strategies and programmes and the support by regional centres . . . Humanitarian organizations bring decades of experience in working with local actors to support local stakeholders to lead adaptation measures to protect their communities against impending climate risks. (International Federation of Red Cross, 2)

The paper goes on to discuss six general strategies to help local communities with Climate Change Adaptation (CCA). These include the coordination of CCA with efforts that address other vulnerabilities of communities (especially in developing countries), such as HIV/AIDS and poverty. Policies should also aim at "capacity building" which will enable local communities "to understand climate risk issues, effectively use available information, develop the necessary institutions and networks, [and] plan and build appropriate CCA actions." A closely related strategy is needed to promote the coordination of CCA measures through "regulatory structures that align the broad range of development activities taking place at national and local levels" (International Federation of Red Cross, 6).

In addition to encouraging local action on CCA, the UN has also integrated climate change issues into its policies on urban planning. The UN Human Settlements Program (UN-Habitat) has a particularly strong concern for how climate change might stress the social order of cities, again, especially in developing countries. Among its many programs is the Climate Change and Cities Initiative (CCCI). Functioning on a global scale, programs such as CCCI strive to build awareness on the paramount role cities and local governments have in addressing climate change. On its Web site, CCCI explains: "Cities have the potential to influence the causes of climate change and they have the solutions to advance climate protection. The success of adaptation critically depends on the availability of necessary resources, not only financial and natural resources, but also knowledge, technical capability, institutional resources and targeted tools" (CCCI). The Web site goes on to highlight many of the same issues regarding capacity building, vulnerabilities, and communication that we just saw with the UNFCCC negotiators:

[The CCCI] seeks to provide support towards the development and implementation of pro-poor and innovative climate change policies and strategies; and to develop tools for enhancing capacities of local governments. Given the tremendous task ahead, UN-HABITAT will work closely with a diverse range of partners within the UN system, governments at all levels, NGOs, communities, institutions of research and higher learning; capacity building and training agencies; land and property organizations, the private sector, among others. (CCCI)

The CCCI currently has projects underway in four pilot cities: Esmeraldas, Ecuador, Kampala, Uganda, Maputo, Mozambique, and Sorsogon, the Philippines. The programs address issues that are specific to each city, including changes in tidal flooding and soil erosion, flooding due to increased precipitation, drought due to lowered precipitation, air pollution due to the expansion of human settlements, and infrastructure damage due to storms and typhoons.

Through these mechanisms, the UN has, ironically, helped to make information for implementing climate change policies more accessible in developing nations than they may be in portions of the United States and a few other developed nations. International communities, however, must be receptive to such strategies. At present the UN works in coordination with international economic entities including the World Bank and other economic development agencies to use stimulus grants to leverage interest from less developed nations. In short, if they implement the ideas of sustainability that may help the climate problem, nations may be eligible for financial assistance.

## FEDERAL INITIATIVES TAKE SHAPE

Grassroots efforts on the domestic front and the model of international efforts have slowly had an impact on federal policies in the United States. Discussions of how American policymakers will contend with climate change are organized primarily around a few specific ideas. Most important, policymakers are currently striving to construct a new "carbon economy" that accounts for the entire lifecycle of energy sources.

Although many nations have stressed the importance of new technology, perhaps the United States has a unique track record in this regard. Since the 19th century, American culture has demonstrated something that Howard Segal calls "technological utopianism," the belief that technology is a particularly efficacious means of bringing about the ideal American society

(Segal, 2005). This tendency has been evident in American efforts of city planning, government reform, and corporate reorganization. New products and businesses have been built on technical breakthroughs such as Franklin's stove, Fulton's steamship, Edison's electric light, and Bell's telephone.

Technological utopianism lives on in America's response to global warming. Many Americans hope to mitigate or adapt to climate change through technological development that not only might sustain the American lifestyle with relatively small modifications, but also might generate new business opportunities. In chapter 4, we discussed some of the more obvious and practical ideas, such as: new or improved energy technology (biofuels and nuclear power), renewable energy sources (wind and solar), improved energy delivery (Smart Grid), more efficient transportation (hybrid cars and mass transit), and revised building codes (improved insulation and efficient appliances).

Beyond these smaller-scale fixes, there are also more ambitious proposals for addressing energy needs and global warming, most of which require significant federal funding for research, development, and demonstration. For instance, chapter 5 discussed the effort to continue the trajectory toward fewer carbon-rich fuels (from coal to oil to natural gas) by moving to hydrogen. Hydrogen burns (via the chemical reaction with oxygen) more cleanly than even methane. Unfortunately, this requires the use of relatively complex fuel cells and high pressures for storage. The hydrogen can be liquefied at low temperatures, but this probably would not be economical. In addition, the gas is highly combustible. Still, there are those who believe that such technical problems can be overcome.

Because the world economy will depend on a significant amount of fossil fuel for the foreseeable future, others have sought to find ways of capturing greenhouse gases from industry smokestacks and storing them through chemical means. One such technology is carbon dioxide trapping and storage, or "carbon sequestration," which moves carbon from the atmosphere into a carbon "sink," or storage area. In the most common version of this idea, $CO_2$ is pumped underground and trapped in rock deposits such as sandstone. Engineers have even suggested pumping $CO_2$ into established oil fields, which, in addition to trapping the greenhouse gas might also increase underground pressure to aid in the recovery of yet more oil. Problems with such proposals include uncertainty about what percentage of the $CO_2$ will remain trapped, the cost of the procedure (which includes transportation from factory to sequestration site), and the inherent danger of $CO_2$ (miners know it as "choke damp," a gas that smothers its victims) (Flannery, 258).

Seeking to trap $CO_2$ in underground sites is an example of geoengineering, the effort to manipulate the Earth's climate, through further human intervention, such that it counteracts the effects of global warming.

Another geoengineering idea was introduced during the 1970s by Mikhail Budyko who suggested that the planet might be cooled by burning sulfur in the stratosphere. Budyko hoped that the resulting haze, somewhat like the volcano emissions noted by Ben Franklin and others, would provide an aerosol canopy that would block part of the sun's radiation, to bring a "global dimming" and slightly cooler global temperatures. About a decade later, Wallace Broecker, of Columbia University, proposed the use of hundreds of planes to dump many tons of sulfur dioxide in the stratosphere to act as an aerosol canopy. The famous American physicist Edward Teller, before his death in 2003, wrote in the *Wall Street Journal*: "Injecting sunlight-scattering particles into the stratosphere appears to be a promising approach. Why not do that?" In fact, there are a number of possibly serious problems with the idea. First, the sulfates dropped in the upper atmosphere will eventually come down with precipitation as acid rain. Second, the aerosols may introduce severe variations in precipitation that may disrupt agriculture. Finally, the concept may prove to be extremely expensive. James Hansen of NASA has criticized it as "incredibly difficult and impractical" (Broad, 2006).

Another geoengineering proposal is to encourage the "biological pump" of the oceans (see chapter 1) by "fertilizing" them with iron dust in order to catalyze the growth of plankton. This is considered an extension of the natural process in which winds carry dust from the Earth's deserts to the oceans, resulting in giant plankton blooms near the shoreline. While the plankton live, they capture $CO_2$ from the water through respiration. After their relatively short lifecycle, they die and a percentage sinks to the bottom of the ocean and therefore removes carbon (in the form of their bodies) from the environment. There are numerous problems with such proposals. Aside from the cost of dispensing tons of iron filings in the oceans, there is considerable doubt about what percentage of the carbon is truly removed from the carbon cycle and transported to the ocean depths (Maslin, 144).

Other ideas seek to avoid a major problem of increased $CO_2$ levels in the oceans. When $CO_2$ dissolves in the water, much of it is chemically transformed into a weak acid, carbonic acid. This leaves the water acidic, which is harmful to many forms of marine life, particularly coral. One way to counteract the acidification is to dump limestone into the water. In addition to the cost of the limestone itself, there also are costs of dispersing it across large surfaces of the ocean. Researchers at Lawrence Livermore National Laboratory (LLNL) pursued an initiative that sought to minimize this second cost by catching the acid at an earlier stage. In the LLNL process, carbon dioxide is first hydrated (exposed to water vapor) as it goes up the smokestacks of power plants, producing a carbonic acid solution. Instead of releasing the carbonic acid directly into deep ocean waters, the acid is first mixed with limestone (Parker, 2004).

Yet another initiative in geoengineering seeks to protect ocean coral against the dangers of rising temperatures and resulting bleaching. Andrew Baker, of the University of Miami's Rosenstiel School of Marine and Atmospheric Science, hopes to take advantage of the natural symbiotic relationship between coral and algae by transplanting a strain of heat-resistant algae in the endangered coral reef areas. Because the alternative strain can withstand higher temperatures, Baker hopes that the corals will adapt to the changing environment by "switching algae" to protect themselves. The idea is being pursued, even though there remain considerable doubts about whether the corals will be able to adapt quickly enough (Velasquez-Manoff, 2009).

Technofixes and geoengineering may have promise for our future response to climate change. Currently, however, there are numerous concerns regarding such efforts. First, applying each of these ideas involves significant technical and economic problems. Beyond these initial difficulties, there is good reason to worry that if the technofixes worked, then they would be regarded as a cure-all for climate difficulties and taken as a carte blanche to "pollute all we want."

## Carbon Accounting

Clearly, the most immediate impact of mitigating climate change concerns ideas of monetizing—or putting a price on—carbon outputs. Accounting for the costs of relying on renewable resources to generate electricity suggests a great boon for attention to the issue of climate change. In the early 21st century, all energy costs have risen at staggering rates. Since 2000, natural gas rates have risen by 80 percent and gasoline has more than doubled in price. Even electricity rates have risen by nearly 40 percent after actually declining during most of the 1990s. Americans have found evidence of these increased costs in all types of related goods as well, particularly agriculture. Across the board, increased fossil fuel prices have trickled into the lives of American consumers and reminded them that we live an energy-intensive lifestyle. Based on cheap fuels, this lifestyle has defined American life for a century. As the prices now rise, the basic cost of living stresses many in the middle class to the breaking point. Also, new attention has been focused on the entire lifecycle of energy resources.

By the early 21st century, the actual price of fossil fuels was evident in both the pollution they produced and the rapidly increasing cost to consumers. Although the economic collapse of 2008–9 gave consumers a break from high-energy costs, prices are expected to soar once global consumption of energy resources has righted itself. If, in the meantime, society does not adapt to use more alternative energy and to make use of much greater efficiency and conservation, the resulting energy prices will probably dwarf those seen before the economic collapse.

And the prices we pay for energy use do not yet include the harmful effects on human health, damage to land from mining, and the environmental degradation caused by global warming, acid rain, and water pollution. When we begin fully accounting for our high-energy existence, then these related environmental and financial costs of pollutants must be included in the cost of "cheap" fossil fuels. Estimates have been made that when energy producers prevent these harmful emissions or otherwise pay for their effects, the cost of fossil fuels doubles—this is referred to as "carbon accounting." With the full accounting of fossil fuel, energy sources and their impacts on human health, the environment, and climate change, alternative energy sources have become mainstream. This full accounting of the price of fossil fuels can be done in a variety of ways. Ideally, the producer of a certain type of energy should be required to pay for its production and all detrimental effects to society and the environment. When this is done, the producer would then pass this cost along to the consumer. The consumer would then have a stronger motivation to choose a low-energy lifestyle.

Even without a complete production-side accounting, government regulation plays an important role in energy accounting with several different methods. The government can provide incentives to those who use renewable energy and purchase lower consuming products. These incentives are nearly always financial in nature, so they do not provide for cleaner air or environment. And while these incentives have not been valued high enough in relation to the health and environmental impact of the use of fossil energy, they have promoted alternative energy and conservation (Black and Flarend).

Another way for the federal government to promote a full accounting of energy production is to establish a carbon tax or carbon trading schemes (such as the cap-and-trade measures described previously). By enacting a carbon tax the government does not stop the emission of carbon dioxide and the accompanying climate change, but it does make those emissions more expensive. The producer of energy that emits carbon dioxide, then, must pass this cost along to the consumer. This is similar to a production-side accounting and encourages energy use from producers who do not emit carbon dioxide.

A third way for the government to be involved is to pass laws to prevent the emission or release of harmful pollutants. This is sometimes called a "command and control" structure by those opposed to it. With this legal requirement, an energy producer must take necessary steps at whatever costs to prevent the harmful pollution. This cost is then passed on to the consumer.

This type of accounting is production-side accounting. If this were done, it would not be necessary for renewable incentives or carbon taxes to be provided. This type of legal requirement to prevent harmful pollution has proven difficult to enact and enforce.

The linchpin for almost every aspect of the "carbon economy" is redefining $CO_2$ to the status of a pollutant, so that the Clean Air Act will apply to it. The Bush Administration resisted efforts of the Environmental Protection Agency (EPA) to recategorize $CO_2$ in this way. Acting on a 2007 ruling of the U.S. Supreme Court (that the Bush Administration disregarded), the Obama Administration is set to regulate carbon dioxide as a pollutant and, thus, usher in the era of the carbon economy. In 2009, Obama's EPA declared $CO_2$ a pollutant and set rules for reducing its harm using the Clean Air Act, which does not require Congressional action. President Obama, as a candidate, pledged to institute a cap-and-trade regulation that would reduce carbon emissions 80 percent by 2050.

Most likely, in coming years the government will use a mix of these accounting schemes that will have the effect of making alternative energy production cost-competitive. As more of these schemes are used to account for additional harmful pollution from the use of fossil fuels, alternative energy will continue to become more cost-effective and perhaps fossil fuels will soon be cost-prohibitive. When this occurs, the demand for action on behalf of global warming will have played a critical role in creating the carbon economy.

### Making Alternatives Primary

Because most alternative energy sources received serious attention in the 1970s, much advancement has been made in their technology. Driven only with government research and development, the application of these energy sources was demonstrated on small scales so that potential problems could be found and solutions could be engineered. After 30 years, many alternative energy technologies have reached a mature level of development waiting for the time when economic and social conditions were right for large-scale application. It is most likely that the American energy future will require a diversified selection of many sources.

The successful deployment of any renewable electricity generation beyond the 20 percent threshold will require the development of a national smart-grid to replace the currently outdated electrical transmission grid (Black and Flarend). The scale and cost of creating a more up-to-date grid are immense and, therefore, require government investment. This infrastructure, then, could be used to more effectively and efficiently move electricity generated by whichever means deemed most appropriate. Each of these trends function to make energy sources previously considered too expensive, much more competitive. The most significant winner may be nuclear power.

The reemergence of nuclear power has been decades in the making. But perhaps most of all, nuclear power plants are cost-competitive, if not cheaper, than the full accounting of fossil fuel power plants. Because nuclear power is now viewed as cost-competitive, industry is now choosing to invest in this technology and it appears that nuclear power will meet a larger portion of our electricity needs in the future. For uranium to be a lasting part of our energy mix, it will become necessary for the science of breeder reactors and reprocessing nuclear waste into new plutonium and thorium fuel to take president over the politics of not wanting to reprocess nuclear waste. This reprocessing of nuclear waste is sometimes called a "closed fuel cycle" to illustrate that fuel is used to make more fuel (Black and Flarend).

Although other nations like Japan, France, and Russia currently reprocess their nuclear waste, the United States has had a policy for more than 30 years not to reprocess nuclear waste. This political policy was adopted in the hopes of stopping the spread of nuclear weapons around the world. As is evidenced by North Korea, Pakistan, India, Israel, and South Africa, however, this policy has failed. Many of the latest reactor designs being pursued internationally assume reprocessing of nuclear waste so that nuclear power can provide energy for centuries more while reducing the amount of high-level waste that must be stored long term.

The most significant change in the energy transition of 2008 was possibly the broadening of production and use of biofuels. In 2006, when President George W. Bush castigated Americans for their "addiction" to oil, he called for the use of alternatives to produce biofuels, including "switch grass." In hindsight, most experts expect that the first decade of the 21st century will appear as a boom for biofuels, including homegrown gasoline and diesel substitutes made from crops like corn, soybeans, and sugarcane. These technologies had been around for a century but were thrust forward as the most effective transitional energy source as humans considered other ways to power transportation. Although most were never intended for use on a massive scale, biofuels became major players in the energy sector with high gas prices. Ethanol production has responded to these factors, increasing from 50 million barrels in 2002 to more than 200 million barrels in 2008.

Declining prices of new technologies will also, inevitably, enhance the viability of more traditional alternative fuel sources, including wind and solar. In the 21st century, European turbine manufacturers have established headquarters in the United States and have helped to make wind one of the fastest-growing sources of energy in the nation. Many experts believe solar is poised to make a similar jump in feasibility. Massive installations planned in California raise questions about the idea that solar power is best deployed on the roofs of houses and businesses (Black and Flarend). Boosting the solar manufacturing base with such large projects is an important step in lowering the cost of solar energy, both for such large commercial projects, as well as smaller distributed roof-top systems, for future generations.

## BUSINESS OPPORTUNITIES OF THE "GREEN ECONOMY"

A by-product of early efforts to create aspects of a "carbon economy" has been a jump in business and employment opportunities, particularly during the economic difficulties of 2009. Although environmentalism on its own right did not stimulate trends in the larger economy, the issue of climate change appears to provide a suitable logic to stimulate existing business owners to shift hiring and expansion patterns as well as to generate entirely new businesses. "Green collar" jobs have become the fastest growing portion of many nations' economies and are now considered a bone fide portion of the employment sector. More then ever before, businesses are actively measuring how climate change initiatives can offer their industries new opportunities for growth. For the business community, climate change represents an impending market shift that will both alter existing markets and create new ones. The difficulty is to assess what timetable changes will follow. Businesses simply need to make certain that they are prepared to fulfill what appear to be significant changes in consumer needs and expectations.

Andrew J. Hoffman, a Professor of Sustainable Enterprise at the University of Michigan's Ross School of Business, expects that this market shift will create a variety of business opportunities, including new demand for emission-reducing technologies, new financial instruments for emissions trading, new mechanisms for global technology transfer, and new opportunities to eliminate sources of greenhouse gases. Hoffman writes: "The shift will affect all companies to varying degrees, and all have a managerial and fiduciary obligation to assess their business exposure and decide whether action is prudent. In short, as the market shift of climate change looms on the business horizon, the argument against action is increasingly harder to make" (Hoffman, 101–118).

Existing companies have taken a variety of approaches. For instance, DuPont has identified the most promising growth markets in the use of biomass feedstocks that can be used to create new bio-based materials such as polymers, fuels, and chemicals. Hoffman continues, "a partnership between DuPont and BP to develop, produce and market a next generation of biofuels. Alcoa has been emphasizing new opportunities in expanding aluminum recycling, which would significantly decrease their emissions of greenhouse gasses. In a very different approach, insurance underwriters such as Swiss Re, are investigating the industry's coverage of natural catastrophes and property loss and how climate change will demand new models" (Hoffman, 101–118).

The new business opportunities presented by climate change, and environmentalism more generally, might be an area that enables a certain amount of pragmatic collaboration between Democrats and Republicans in an otherwise polarized political culture. This is the hope of Newt Gingrich, one of the main figures of the "Republican Revolution" of 1994, and one of the authors of the famous *Contract with America*. In 2007 he published a book with Terry L. Maple, *A Contract with the Earth*, in which he argues for a "mainstream, nonpartisan approach to environmental problem solving" that draws on the traditional American strengths of technological innovation and free enterprise. Such "entrepreneurial environmentalism" requires the country to expand its workforce of environmental engineers and technologically competent entrepreneurs. Although Gingrich and Maple are doubtful of reliance on government regulation, they are enthusiastic about policies in which the government facilitates innovation and economic development, including funding for education and peer-review grants and prizes, and government incentives to private enterprise in the form of tax credits and subsidies. They write:

Government, at all levels, should be a facilitator for entrepreneurial, private-sector innovations and the formation of private-public environmental partnerships, supporting and not suppressing the creativity of entrepreneurial environmentalists. The resources of government should be consistently applied to reduce red tape and facilitate progress. Some regulation will always be necessary, but it should be limited, focused, and reasonable to liberate the potential of market innovation. (Gingrich and Maple, 13)

Gingrich and Maple include in their book a number of examples of companies that have already shown success in turning environmental problems into profitable businesses, including the collection of methane gas from landfill and the production of ethanol from cow manure. Commenting on the relatively recent problem of discarded electronic devices (computers, cell phones, games, and so on), Gingrich and Maple comment with characteristic optimism: "Still, a clever entrepreneur is going to figure out how to turn this problem into a 20- to 50-million ton business opportunity" (Gingrich and Maple, 92).

Gingrich and Maple's view of global warming follows from their general argument regarding environmentalism and stresses examples of companies that have already responded to the challenge of reducing carbon emissions. Oil companies such as Shell Oil and Beyond Petroleum (formerly known as

British Petroleum) have made public statements that the future expansion of energy infrastructure must go hand–in-hand with a change in carbon profile, and have already invested in solar and wind power. Gingrich and Maple also look to enlightened philanthropy as a path to innovation and speak glowingly of Richard Branson's investment group Virgin Fuels, which will devote $3 billion over the next decade for research on renewable energy sources. Finally, they point enthusiastically to the foundation of coalitions like the U.S. Climate Action Partnership (USCAP), the membership of which includes many large American companies such as Alcoa, General Electric, and PepsiCo.

An entrepreneurial environmentalism such as Gingrich and Maple's is, not surprisingly, suspicious of the visions of catastrophe or doomsday that they feel have been encouraged by certain environmentalists. This concern is particularly acute regarding the issue of global warming. Although Gingrich and Maple support strong research on climate change, they worry that hasty remarks made by certain scientists will be picked up as sensationalist headlines in the media and eventually lead to restrictive legislation. To combat this, Gingrich and Maple have issued a call for openness in gathering and analyzing climate data:

The maintenance and continuous improvement of a worldwide data system should be a high priority. Furthermore, the data should be available to everyone. Scientific debate and dissent should be encouraged in the pursuit of a thorough and comprehensive understanding of the complexities of our environmental systems. (Gingrich and Maple, 190)

Certainly, their remarks are prescient in light of the controversies over the hacked e-mails released from the University of East Anglia in late 2009 (see Epilogue). It is easier to be critical about their unclear stance regarding government regulation. Although Gingrich and Maple are not enthusiastic about the Kyoto Protocol, which they view as "clearly flawed," they do agree that "something should be done to achieve sensible reductions in greenhouse gas emissions" (Gingrich and Maple, 34). It is not clear how far they will go from government as facilitator toward the government as regulator (involving a cap and trade policy, which is supported by the USCAP). This general issue—how much companies will be able to adapt to challenges associated with global warming and how much the federal government will be forced to intervene—is highlighted by our final example involving automobile manufacture.

## Greening Our Rides

One of the first demonstrations of the seriousness of our energy change has been the shift in automobiles demanded by the U.S. consumer starting in 2005 and reaching a fever pitch by 2008. The prolonged cost increase in petroleum to the $4 per gallon range irreparably altered the auto marketplace and demonstrated just how much influence consumers could have on the auto industry. Toyota and Honda led the way by making hybrid vehicles widely available. As Americans' love affair with large vehicles gave way to heightened economic concerns, drivers chose smaller vehicles. American manufacturers were left near complete ruin as a result of their emphasis on manufacturing larger vehicles including SUVs and full-size pickup trucks.

In the case of America's Big 3—those companies that delivered 10,000-pound large vehicles for middle-class consumers—entire plants dedicated to manufacturing SUVs and trucks have been shut down or shifted to making smaller cars. As reported by Bill Vlasic in the *New York Times,* "the biggest losers in the market are the big pickups and SUVs that Ford and its domestic rivals," General Motors and Chrysler, rely on for much of their profits . . . 'We saw a real change in the industry demand in pickups and S.U.V.'s in the first two weeks of May [2008],' Ford's chief executive, Alan R. Mulally, said. . . . "It seems to us we reached a tipping point'."

Overall, the momentum in small-car sales is outpacing industry growth worldwide, the automaker said. In a January 2008 press release, Ford noted that "globally, small car sales are expected to grow from 23 million in 2002 to an estimated 38 million in 2012. . . . Driving the growth in the North American market is a group of young consumers age 13 to 28 years—dubbed "Millennials." Today, this group numbers approximately 1 billion worldwide and will represent 28 percent of the total U.S. population by 2010." These consumers, who have grown up entirely in a world of high-priced gasoline, realize that alternatives are a must ("Ford Reveals Small-Car Vision for North America").

In addition to shrinking the types of vehicles composing its fleet, manufacturers also stepped up efforts to create commercially available hybrids and alternatively fueled vehicles. First in 2006 and again in 2008, Honda and Toyota were being forced to use waiting lists for their commercially available hybrids. Each American company advertised hybrid models, but few of them actually made it to the road. Instead, GM and Ford each strategically elected to develop plug-in, all-electric vehicles that are supposed to be available commercially by 2010, and it is now known that these will only be available in limited quantities. They also made their fleet appear more green by selling vehicles able to use more biofuels.

The larger economic collapse of 2008–9 took a bad situation for American manufacturers and made it grave. At the time of this writing, only Ford Motor Company has survived without government assistance. And the federal government was poised to take over American auto manufacturing to stem the tide of job losses particularly in Midwest states. The long-term future of worldwide automobile manufacturing will probably depend on the speed with which companies can emerge from the current economic downturn with the successful mass-production of mid-transition alternative-fueled vehicles.

## INNOVATION AND BACKLASH

This chapter traces a remarkable shift in American thought and policy over a brief period. Since 1970, global warming has gone from a speculative scientific hypothesis that went virtually unnoticed by the general public to an established scientific consensus, encouraging grassroots policy initiatives, innovative technological developments, new business opportunities and, eventually, significant policy changes at the federal level. All this reached something of an apex when, on June 26, 2009, the House of Representatives approved the American Clean Energy and Security Act of 2009—also known as the Waxman-Markey Bill—by the narrow margin of 219–212. This historic bill, if it is eventually signed into law, would subsidize technological innovation toward lower-carbon technology and institute a national cap and trade system.

Such rapid cultural and political shifts, however, are never without difficulty. Less than two months later, a *New York Times* headline read: "Oil Industry Backs Protests of Emissions Bill." Describing an event in Houston, Texas organized by a group called Energy Citizens, which is backed by the American Petroleum Institute, the article explained how employees of oil companies who work in Houston were bused from their workplaces:

This [event] was expected to be the first of a series of about 20 rallies planned for Southern and oil-producing states to organize resistance to proposed legislation that would set a limit on emissions of heat-trapping gases, requiring many companies to buy emission permits. Participants described the system as an energy tax that would undermine the economy of Houston, the nation's energy capital.

Mentions of the legislation, which narrowly passed the House in June, drew boos, but most of the rally was festive. A high school marching band played, hot dogs and hamburgers were served, a video featuring the country star Trace Adkins was shown, and hundreds of people wore yellow T-shirts with slogans like "Create American Jobs Don't Export Them" and "I'll Pass on $4 Gas." (Krauss and Mouawad)

Although national polls show that a majority of Americans support efforts to tackle climate change, it is also true that Americans remain suspicious of top-down solutions from the federal government. Added to this, opposition to climate change legislation from energy-intensive industries has become more vigorous in 2009–10.

The Waxman-Markey Bill that passed the House seeks to reduce greenhouse gases in the United States by 83 percent by 2050. To ensure passage of the bill, sponsors doled out 85 percent of the carbon permits for free. The power sector received 35.5 percent of the free allowances, but petroleum refiners received only 2.25 percent. This is only one of the factors stirring the current backlash. In short, establishing the need for initiatives to fight climate change may have been the easy part. As policymakers work to construct federal and international initiatives that actually put these plans into practice, they are sure to run into some of the most entrenched interests of the U.S. economy. The success of many regional, state, and local initiatives may wind up being models for the future projects on a much larger scale.

# Epilogue

# Integrating Global Warming into Federal Action

So we have a choice to make. We can remain one of the world's leading importers of foreign oil, or we can make the investments that would allow us to become the world's leading exporter of renewable energy. We can let climate change continue to go unchecked, or we can help stop it. We can let the jobs of tomorrow be created abroad, or we can create those jobs right here in America and lay the foundation for lasting prosperity.

—President Barack Obama, March 19, 2009

When Americans elected Barack H. Obama its 44th President, the nation appeared to have turned a corner. Although throughout the campaign candidate Obama was not as forceful as many environmentally-minded voters may have preferred in his statements about climate change and new energy, his administration has taken a relatively strong stand in pursuing a holistic, integrated planning of the nation's energy future *that takes climate change into consideration.* In short, the administration has accepted the reality of climate change and given serious consideration to the issue of mitigation. Where the U.S. Senate voted 95–0 to spurn the Kyoto talks in 1997, Congress in 2010 is considering the Waxman-Markey bill that aims to cap and reduce greenhouse-gas emissions. This moment of concerted, international effort to solve this common problem allows Americans the opportunity to consider the primary question at hand: What would energy management look like if a nation accepted climate change and sought to truly move toward modification?

In *Hot, Flat, and Crowded, New York Times* columnist Tom Friedman designated 2007–8 the first year of the "Energy Climate Era" (ECE), an era when

"We now understand that these fossil fuels are exhaustible, increasingly expensive, and politically, ecologically, and climatically toxic" (Friedman, 38). In the past, he writes, "we wanted everyone to be converted to the American way of life, although we never really thought about the implications. Well, now we know. We know that in the Energy Climate Era, if all the world's people start to live like us . . . it would herald a climate and biodiversity disaster" (76). Ironically, change has come quicker to other nations and the United States finds itself lagging behind—even trying to imitate other nations that have more readily modeled their economies on Friedman's ECE.

Although the United States trails other nations because of its sluggish start, the world's largest consumer of energy now appears to have joined many other developed nations in an energy transition away from fossil fuels and toward the greener methods that Friedman mentions. Much of this technology grew from flashpoints very reminiscent of previous energy transitions. In the Netherlands, for instance, the government has invested more than $80 million to restore some of the 1,040 older mills already in existence. Discussed briefly in chapter 3, the cradle of energy from wind has been born again in a new era when many of the ancient mills have been retrofitted to generate electricity instead of to grind grain. In addition, the government has constructed one large-scale wind farm off the coast and has plans for others. Making the Netherlands' adoption of alternative power easier, of course, is the nation's small population, size, and, commensurately, footprint. Such changes are more complicated in nations that have allowed themselves to grow more dependent on fossil fuels. However, the American wind initiative is growing, too, helping to lead a new era of alternative energy toward becoming a reality.

Clearly, though, global warming forms a basic logic in creating the tipping point for massive changes such as energy transitions. As the first decade of the 21st century comes to a close, policy thinkers have integrated climate change into our nation's future, in the form of "carbon accounting" and other mechanisms. In short, knowledge of climate change has led policymakers to think about the entire life cycle of energy sources to account for the problems or pollution that they leave behind. Similar to a surtax, these costs must be added to the price of an energy source. In this new paradigm, the American lifestyle has become unsustainable as "cheap" fuels such as coal and petroleum have been recalibrated. We hear echoes of Carter's "malaise speech," discussed in chapter 4. Now, however, after more than a decade of debate over whether or not humans had any responsibility to act to stop climate change, it has become evident that action needs to come from all over the ideological spectrum—liberal or conservative, Republican or Democrat. It is particularly important in the United States because we have seen our nation's competitive advantage in so many things slip.

The United States has been slower than the Netherlands and many European nations to create effective government stimuli for the development of wind power and other alternative energy. Inspired in some ways by modern environmentalism and in others by new business opportunities, green power options moved to the mainstream in the 21st century, including incorporation into economic stimulus initiatives of 2009.

How does a leader guide an energy transition forward while governing the current patterns with which Americans live? Although each candidate for U.S. President in 2008 discussed initiatives in this area, actually governing with an eye towards sustainability is much more problematic. In the late 1970s, President Jimmy Carter demonstrated the difficulty of the Oval Office attempting to lead technological innovation. It appears that the Obama Administration has adopted a more integrated approach than that of Carter or any other U.S. president. Such initiatives, though, succeed or fail based on the public reaction to them.

The Obama initiatives take on a diversity of issues but share a common approach and ethic, including these: more than $60 billion contained in the American Recovery and Reinvestment Act for purposes such as developing a smarter grid that will move renewable energy from the rural places it is produced to the cities where it is needed most and 40 million smart meters to better measure the energy used in American homes, weatherizing homes, constructing green federal buildings, billions of stimulus dollars for state and local renewable energy and energy efficiency efforts and green job training programs, and $2 billion in competitive grants to develop the next generation of batteries to store energy, primarily for vehicles. Finally, with a focus on transportation, the administration for the first time in a decade raised fuel economy standards so that Model Year 2011 cars and trucks will get better mileage, and the successful "Cash for Clunkers" took thousands of less efficient cars off the roads and helped consumers replace them with more efficient models.

Additionally, the United States has resumed active participation in the global discussion on climate change. As discussed in chapter 5, there was good reason for optimism before the 15th meeting of the Council of the Parties of the UNFCCC (COP-15) in Copenhagen. In preparation for this meeting, President Obama joined other North American leaders in August 2009 in releasing a memorandum titled "North American Leaders' Declaration on Climate Change and Clean Energy," which began with the words:

We, the leaders of North America reaffirm the urgency and necessity of taking aggressive action on climate change. We stress that the experience developed during the last 15 years in the North American region on environmental cooperation, sustainable development, and clean energy research, development, and deployment constitutes a valuable platform for climate change action, and we resolve to make use of the opportunities offered by existing bilateral and trilateral institutions.

We recognize the broad scientific view that the increase in global average temperature above pre-industrial levels ought not to exceed 2 degrees C, we support a global goal of reducing global emissions by at least 50% compared to 1990 or more recent years by 2050, with developed countries reducing emissions by at least 80% compared to 1990 or more recent years by 2050.

We share a vision for a low-carbon North America, which we believe will strengthen the political momentum behind a successful outcome at the 15th Conference of the Parties to the UNFCCC meeting this December, and support our national and global efforts to combat climate change. (North American Leaders' Declaration)

The balance of the Declaration agrees to pursue a number of general goals, including the support of mitigation and adaptation actions, the development of methods of measuring, reporting, and verifying emissions reductions, and the collaboration on low-carbon technologies. The Declaration highlighted a number of key sectors where emission reductions may be found, including protection of forests and wetlands, and a reduction of GHG emissions in the transportation and oil and gas sectors.

Obama's approach seems to be clear. However, on the issue of climate change, the international stage and on the American domestic scene there is still significant discord and dissention. Can a political leader manufacture consensus and simply move the issue forward? No.

Two disappointments, coming in the closing months of 2009, will no doubt limit the ability for the federal government to act in the immediate future. The first disappointment came with the 15th Conference of the Parties (COP-15) of the UNFCCC in Copenhagen, which was attended by representatives from over 190 countries. Its goal was to establish a new international agreement which could take force when the Kyoto Protocol expires in 2012. A series of meetings during 2009 before the conference gave a hopeful picture, as a number of nations agreed to emissions-reduction targets, in the event that a binding agreement could be reached. Unfortunately, the conference negotiations proved difficult for many of the same reasons mentioned in chapter 5 in our discussion of debate over the Kyoto Protocol (including fairness of the agreement between the developed and undeveloped nations and difficulties with cap and trade schemes). In the end, the conference produced little more than a statement of hopes, one that underlined the importance of the issue of climate change, a statement that did not contain any legally binding caps or targets.

Reactions to this outcome varied. Some chose to stress that Copenhagen had been a failure and that this was a great setback for the mitigation of climate change. A few months after the meeting, the executive secretary of the UNFCCC, and one of the most visible negotiators at Copenhagen, Yvo de Boer resigned his post. Elliot Diringer of the Pew Center on Global Climate Change noted that, "Rightly or wrongly, Yvo is associated in many minds with the perceived failure of Copenhagen and no longer has the confidence of parties" (Samuelsohn, 2010).

Others, including James Hansen, felt that Copenhagen had to fail because the international community was not yet ready to reach an efficacious agreement. As a supporter of a carbon tax or fee, Hansen is dubious of the cap and trade approaches now favored by the UNFCCC: "This is analagous to the indulgences that the Catholic Church sold in the middle ages. The bishops collected lots of money and the sinners got redemption. Both parties liked that arrangement despite its absurdity. That is exactly what's happening. We've got the developed countries who want to continue more or less business as usual and then these developing countries who want money and that is what they can get through offsets [sold through the carbon markets]" (Goldenburg, 2009).

Still others chose to stress that although Copenhagen did not pass any legally binding agreement, it did reaffirm a number of specific issues that could be addressed at the next meeting (at COP-16, in Cancun, Mexico at the end of 2010). Most generally encouraging was the renewed acknowledgement of the importance of the Kyoto agreement itself and the need for a successor. Many judgments involved seeing the glass half full or half empty. On the plus side, some observers noted that a number of key countries (such as China and India) proposed targets for themselves for the first time. On the negative side, critics could easily claim that pledges like this are easy in an agreement that is not legally binding. Other observers saw progress on specific policies such as "reduce emissions from deforestation and degradation," often referred to as REDD. This approach would involve payments by the developed countries to developing countries in exchange for the developing countries "to reduce emissions from forested lands and invest in low-carbon paths to sustainable development." Many critics, however, are unhappy with this policy, some claiming that it would encourage unscrupulous short-term gain (such as leveling older forests and replanting them with new growth to get carbon credits) (UN-REDD Programme).

A second major disappointment of late 2009 came as the aftermath of a computer break-in at the Climate Research Unit (CRU) at the University of East Anglia. At some point in 2009, someone gained access to a server used by the CRU and removed thousands of e-mails and documents exchanged by CRU scientists for the last decade. In November, these documents were posted on the Internet in an apparent attempt to embarrass the researchers.

And, indeed, some of the messages were embarrassing. Some showed the climate researchers in private conversation being catty and callous in their discussions of their critics and of climate change skeptics.

Most damming, however, were messages in which researchers appeared to be talking about sloppy or dishonest data analysis techniques. The most famous message was from Phil Jones of CRU to researchers in the United States, including Michael Mann, Raymond Bradley, and Malcolm Hughes, the authors of the famous "hockey stick" graph of the temperature changes of the last 1,000 years. Commenting on his own research with temperature records, Jones said "I've just completed Mike's Nature trick of adding in the real temps to each series for the last 20 years (i.e., from 1981 onward) and from 1961 for Keith's to hide the decline." Needless to say, many climate skeptics and political observers seized on quotations like this one to prove that something was wrong with the science of climatology and global warming.

Perhaps the first thing to note about this incident is that it is part of a unique development in which an entire community of scientists, holding a certain consensus position, is attacked for a variety of political and philosophical reasons. Spencer Weart, an experienced historian of science who has written a fundamental book on the history of global warming research, commented about the hacked e-mails:

It's a symptom of something entirely new in the history of science: Aside from crackpots who complain that a conspiracy is suppressing their personal discoveries, we've never before seen a set of people accuse an entire community of scientists of deliberate deception and other professional malfeasance. . . . In blogs, talk radio and other new media, we are told that the warnings about future global warming issued by the national science academies, scientific societies, and governments of all the leading nations are not only mistaken, but based on a hoax, indeed a conspiracy that must involve thousands of respected researchers. Extraordinary and, frankly, weird. Climate scientists are naturally upset, exasperated, and sometimes goaded into intemperate responses; but that was already easy to see in their blogs and other writings. (Weart interview)

Indeed, the scientists' response to charges by politicians and the popular media has been that their words have been quoted out of context. When seen in the context of a private conversation about their research, their choice of words "trick" and "hide" does not show wrongdoing. A posting on the Real-Climate Web site (which is run in part by Mann and Bradley) explained:

Scientists often use the term "trick" to refer to "a good way to deal with a problem," rather than something that is "secret," and so there is nothing problematic in this at all. As for the "decline," it is well known that Keith Briffa's maximum latewood tree ring density proxy diverges from the temperature records after 1960 . . . and has been discussed in the literature since Briffa et al., in *Nature* in 1998 [*Nature*, 391, 678–682]. Those authors have always recommended not using the post-1960 part of their reconstruction, and so while "hiding" is probably a poor choice of words (since it is "hidden" in plain sight), not using the data in the plot is completely appropriate. ("The CRU Hack")

Other observers point out, however, that there are further lessons to be learned from the incident. James Hansen, when asked if the hacked e-mails put the case for anthropogenic climate change into question, responded unequivocally: "No, they have no effect on the science. The evidence for human-made climate change is overwhelming." However, when asked if the e-mails indicate any unethical behavior, Hansen suggested some criticism of the climate community:

They indicate poor judgment in specific cases. First, the data behind any analysis should be made publicly available. Second, rather than trying so hard to prohibit publication of shoddy science, which is impossible, it is better that reviews, such as by IPCC and the National Academy of Sciences, summarize the full range of opinions and explain clearly the basis of the scientific assessment. (Hansen)

Other scientists seem to agree with Hansen. One is Judith Curry, the Chair of the School of Earth and Atmospheric Sciences at the Georgia Institute of Technology, who suggests that the climate community became overly insular as a result of "circling the wagons" against the deniers. She also notes that climate scientists often ignored independent scientists (such as Steve McIntyre, who runs a Web site critical of the climate community, climateaudit.org) requesting the release of climate data for independent confirmation. In an attempt to mend the loss of public trust brought on by the incident, Curry has called for greater openness:

Given the growing policy relevance of climate data, increasingly higher standards must be applied. . . . Scientists claim they would never get any research done if they had to continuously respond to skeptics. The counter to that argument is to make all of your data, metadata and code openly available. Doing this would keep molehills from growing into mountains. (Pearce, 2010)

The U.S. government certainly is impeded in acting on climate change in the face of disagreement on the international stage and of distrust or doubt in the general public. It will not be easy to rebuild the fragile international political consensus that existed for the Kyoto Protocol and to maintain the concern and interest of the general public that polls indicate was significant during the last decade.

Change of this magnitude requires united efforts by policymakers in each party to create initiatives for the entire nation that are based on the local and regional efforts to combat climate change already underway. Overcoming political differences to enhance the future viability of the nation requires that the voting public also accept that the status quo must change. Finally, as consumers, these voters can help to make the greatest of changes by supporting and purchasing the new innovations being developed whenever possible.

The American public is one of the great reasons for hope in this situation. Since Americans first considered energy conservation as a portion of their lifestyle in the 1970s, modern environmentalism and new business opportunities have encouraged an entirely new genre of consumption referred to as "green consumerism." In fact, across the board, mass consumption contains a thread of greenness—conservation thought—that seems to be offering a correction or balance to the high-energy lifestyle in the mid-20th century. History has taught us that such shifts in lifestyle do not fare well when presented to Americans from the top down. Instead, we now operate in an information era in which well-informed consumers might steer producers toward more sustainable and, often, economical uses of energy. It seems reasonable to hope that the second decade of the 21st century might offer the perfect storm of insightful leadership and willing, educated consumers.

Friedman places his hope for the American future in the construction of a "green economy," in which environmental planning is "no longer a fad, green is no longer a boutique statement, green is no longer something you do to be good and hope that it pays off in ten years. Green is the way you grow, build, design, manufacture, work, and live" because it is, simply, smarter and better. "That," he concludes, "is the huge transition we are just beginning to see. Green is going from boutique to better, from a choice to a necessity, from a fad to a strategy to win, from an insoluble problem to a great opportunity" (172). The energy frontier in an era that internalizes the demands of climate change calls for flexibility and adaptation.

A one-size-fits-all energy strategy neither can nor should be mandated from the U.S. federal government. Nor can society wait for a perfect solution to present itself as the path to a new energy future. By waiting for a perfect solution, America will fail to move forward and will ultimately rely on technologies developed in nations that have more actively pursued alternative sources of energy. The successful freeway to America's energy future will have many lanes representing a variety of energy sources including even the clean use of remaining fossil fuels. In the case of climate legislation, Kyoto, most observers agree, marked an imperfect start to an international discourse on the issue. As the dialogue continues at COP-16, Americans must afford our leaders the flexibility they will require in negotiating with other nations, including carbon taxes in addition to cap and trade. Most experts agree that such American support would greatly assist any policy receiving the support of industrializing powers, particularly India and China.

By early 2010, the Obama Administration had demonstrated a much more holistic view of energy than any of its predecessors. Most important, the considerations of climate change were brought into the mix and included in the conception of an overall federal approach to energy. The time is ripe for American consumers to do the same and to help their nation take important steps toward carbon neutrality. Whether the American public and their representatives can take these steps depends on a reevaluation and reinvention of the American way of life in the face of unavoidable uncertainty. Spencer Weart put this well when he wrote:

Faced with scientists who publish warnings, the public's natural response is to ask them for definitive guidance. When the scientists fail to say for certain what will happen, politicians habitually tell them to go back and do more research. In the case of climate, waiting for absolute certainty would mean waiting forever. . . .

If there is even a small risk that your house will burn down, you will take care to install smoke alarms and buy insurance. We can scarcely do less for the well-being of our society and the planet's ecosystems. Thus the only useful discussion is over what measures are worth their cost. (Weart, "A Personal Note.")

# Selected Resources

Andrews, Richard N. L. *Managing the Environment, Managing Ourselves.* New Haven: Yale University Press, 1999.

Athansiou, Tom. *Divided Planet: The Ecology of Rich and Poor.* Athens: University of Georgia Press, 1998.

Begley, Sharon. "It's Not the Kind of Thing Where You Can Compromise." *Newsweek,* November 24, 2009. http://www.newsweek.com/id/224178.

Black, Brian. *Petrolia: The Landscape of America's First Oil Boom.* Baltimore, Md.: Johns Hopkins University Press, 2000.

Black, Brian, and Richard Flarend. *Alternative Energy.* Santa Barbara, Calif.: ABC-CLIO, 2010.

Boli, J., and G. Thomas. *Constructing World Culture: International Nongovernmental Organizations Since 1875.* Stanford, Calif.: Stanford University Press, 1999.

Bowen, Mark. *Censoring Science: Inside the Political Attack on Dr. James Hansen and the Truth of Global Warming.* New York: Plume, 2009.

Bradsher, Keith. *High and Mighty: SUVs: The World's Most Dangerous Vehicles and How They Got That Way.* New York: Public Affairs, 2002.

Brennan, Timothy J., et al. *A Shock to the System—Restructuring America's Electricity Industry.* Washington, D.C.: Resources for the Future, 1996.

Brinkley, Douglas. *Wheels for the World: Henry Ford, His Company and a Century of Progress.* New York: Viking, 2003.

Broad, William J. "How to Cool a Planet (Maybe)." *New York Times,* June 27, 2006. http://www.nytimes.com/2006/06/27/science/earth/27cool.html.

Broder, John M. "Climate Change Seen as Threat to U.S. Security." *New York Times,* August 8, 2009. http://www.nytimes.com/2009/08/09/science/earth/09climate.html.

Broder, J. M. "Rule to Expand Mountaintop Coal Mining," *New York Times,* August 23, 2007. Late Edition—Final, Section A, Page 1.

Brower, Michael. *Cool Energy: Renewable Solutions to Environmental Problems,* rev. ed. Cambridge, Mass.: MIT Press, 1992.

Bruegmann, R. *Sprawl: A Compact History.* Chicago: University of Chicago Press, 2005.

Bryant, Edward. *Climate Process and Change.* Cambridge: Cambridge University, 1997.

Buckley, Geoffrey L. *Extracting Appalachia: Images of the Consolidation Coal Company, 1910–1945.* Akron: Ohio University Press, 2004.

Burroughs, William James. *Climate Change: A Multidisciplinary Approach.* 2nd ed. Cambridge: Cambridge University, 2007.

Cantelon, Philip, and Robert C. Williams. *Crisis Contained: Department of Energy at Three Mile Island.* Carbondale: Southern Illinois University Press, 1982.

Carson, Rachel. *Silent Spring.* New York: Mariner Books, 2002.

Chambliss, Lauren. "Addressing Global Climate Change at the Local Level: A CALS-led Team Shows how Tompkins County can cut its 'Carbon Footprint' by Two-Thirds." *eCALSconnect* 14–2 (December 2007). http://www.cals.cornell.edu/cals/public/comm/pubs/ecalsconnect/vol14-2/features/carbon-footprint.cfm.

Chernow, Ron. *Titan: The Life of John D. Rockefeller, Sr.* New York: Random House, 1998.

Christianson, Gale E. *Greenhouse: The 200-Year Story of Global Warming.* New York: Walker and Co., 1999.

Colignon, Richard A. *Power Plays.* Albany: SUNY Press, 1997.

Cooper, Gail. *Air-Conditioning America.* Baltimore: Johns Hopkins, 2002.

Cotton, William R., and Roger A. Pielke Sr. *Human Impacts on Weather and Climate,* 2nd ed. Cambridge: Cambridge University Press, 2007.

Cronon, William. *Changes in the Land.* New York: Norton, 1991a.

Cronon, William. *Nature's Metropolis.* New York: Norton, 1991b.

Crosby, Alfred. *Children of the Sun: A History of Humanity's Unappeasable Appetite for Energy.* New York: Norton, 2006.

Cutright, Paul. *Theodore Roosevelt: The Making of a Conservationist.* Urbana: University of Illinois Press, 1985.

Darst, Robert G. *Smokestack Diplomacy: Cooperation and Conflict in East-West Environmental Politics.* Cambridge, Mass.: MIT Press, 2001.

Daumas, Maurice, ed. *A History of Technology and Invention,* vol. 3: *The Expansion of Mechanization, 1450–1725.* New York: Crown, 1969.

Davis, Devra. *When Smoke Ran Like Water: Tales of Environmental Deception and the Battle Against Pollution.* New York: Basic Books, 2002.

Diamond, Jared. *Guns, Germs, and Steel: The Fates of Human Societies.* New York: Norton Books, 1997.

Diamond, Jared. *Collapse: How Societies Choose to Fail of Succeed.* New York: Penguin Books, 2005.

Doyle, Jack. *Taken for a Ride: Detroit's Big Three and the Politics of Air Pollution.* New York: Four Walls Eight Windows, 2000.

Eichstaedt, P. H. *If You Poison Us: Uranium and Native Americans.* Sante Fe, NM: Crane Books, 1994.

Fagan, Brian. *The Long Summer: How Climate Changed Civilization.* New York: Basic Books, 2004.

Flannery, Tim. *The Weather Makers: How Man Is Changing Climate.* New York: Grove Press, 2005.

Fleming, James Rodger. *Historical Perspectives on Climate Change.* New York: Oxford University Press, 1998.

Flink, James J. *The Automobile Age.* Cambridge: MIT Press, 1990.

Freedman, Andrew. "Science Historian Reacts to Hacked Climate E-mails." *Washington Post,* November 23, 2009. http://voices.washingtonpost.com/capitalweather gang/2009/11/perspective_on_a_climate_scien.html.

Freese, Barbara. *Coal: A Human History.* New York: Perseus, 2003.

Friedman, Thomas. *Hot, Flat, and Crowded: Whey We Need a Green Revolution—and How It Can Renew America.* New York: Farrar, Strauss, and Giroux, 2008.

Gardner, J. S., and P. Sainato. "Mountaintop Mining and Sustainable Development in Appalachia," *Mining Engineering* (March 2007): 48–55.

Garwin, Richard L., and Georges Charpak. *Megawatts and Megatons: A Turning Point in the Nuclear Age.* New York: Knopf, 2001.

Gelbspan, Ross. *The Heat Is On: The Climate Crisis.* Reading, Mass.: Perseus Books, 1995.

Gingrich, Newt, and Terry L. Maple. *A Contract with the Earth.* Baltimore: Johns Hopkins University Press, 2007.

Goldenberg, Suzanne. "Copenhagen Climate Change Talks Must Fail, says top Scientist." *The Guardian,* December 2, 2009. http://www.guardian.co.uk/ environment/2009/dec/02/copenhagen-climate-change-james-hansen.

Gordon, Richard, and Peter VanDorn. *Two Cheers for The 1872 Mining Law.* Washington, D.C.: CATO Institute, April 1998.

Gordon, Robert B., and Patrick M. Malone. *The Texture of Industry.* New York: Oxford, 1994.

Gore, Al. *An Inconvenient Truth: The Planetary Emergency of Global Warming and What We Can Do About it.* New York: Rodale Books, 2006.

Gorman, Hugh. *Redefining Efficiency: Pollution Concerns.* Akron, Oh.: University of Akron Press, 2001.

Gottlieb, Robert. *Forcing the Spring: The Transformation of the American Environmental Movement.* Washington, D.C.: Island Press, 1993.

Gutfreund, Owen D. *20th Century Sprawl: Highways and the Reshaping of the American Landscape.* New York: Oxford University Press, 2005.

Hampton, Wilborn. *Meltdown: A Race against Nuclear Disaster at Three Mile Island: A Reporter's Story.* Cambridge, Mass.: Candlewick Press, 2001.

Hancock, Paul L., and Brian J. Skinner. *The Oxford Companion to the Earth.* Oxford: Oxford University Press, 2000.

Hardin, Garrett. "The Tragedy of the Commons." *Science* 162 (1968): 1243–48.

Hays, Samuel P. *Beauty, Health, and Permanence: Environmental Politics in the United States, 1955–85.* New York: Cambridge University Press, 1993.

Hays, Samuel P. *Conservation and the Gospel of Efficiency.* Pittsburgh: University of Pittsburgh Press, 1999.

Hoffman, A. "The Coming Market Shift: Climate Change and Business Strategy." In *Cut Carbon, Grow Profits: Business Strategies for Managing Climate Change and Sustainability,* ed. K. Tang and R. Yoeh. London: Middlesex University Press, London, 2007.

Holton, Gerald. *Thematic Origins of Scientific Thought: Kepler to Einstein,* rev. ed. Cambridge, Mass.: Harvard University Press, 1988.

Horowitz, Daniel, *Jimmy Carter and the Energy Crisis of the 1970s.* New York: St. Martin's Press, 2005.

Houghton, John. *Global Warming: The Complete Briefing,* 4th ed. Cambridge: Cambridge University, 2009.

Hughes, Thomas. *Networks of Power: Electrification in Western Society, 1880–1930.* Baltimore: Johns Hopkins University Press, 1983.

Hughes, Thomas. *American Genesis.* New York: Penguin, 1989.

Hunter, Louis C., and Bryant Lynwood. *A History of Industrial Power in the United States, 1780–1930,* vol. 3: *The Transmission of Power.* Cambridge, Mass.: MIT Press, 1991.

Hurley, Andrew. *Environmental Inequalities: Class, Race, and Industrial Pollution in Gary, Indiana, 1945–1980.* Chapel Hill: University of North Carolina Press, 1995.

IPCC. *Climate Change: The IPCC Scientific Assessment.* Cambridge: Cambridge University Press, 1991.

IPCC. *Climate Change 1992: The Supplementary Report to the IPCC Scientific Assessment.* Cambridge University Press, 1993.

IPCC. *Climate Change 1995: The Science of Climate Change Contribution of Working Group I to the Second Assessment Report of the Intergovernmental Panel on Climate Change.* Cambridge: Cambridge University Press, 1996.

IPCC. *Climate Change 2001: The Scientific Basis. Contribution of Working Group I to the Third Assessment Report of the Intergovernmental Panel on Climate Change.* Cambridge: Cambridge University Press, 2001a.

IPCC. *Climate Change 2001: Synthesis Report. Contribution of Working Groups I, II, and III to the Third Assessment Report of the Intergovernmental Panel on Climate Change.* Cambridge: Cambridge University Press, 2001b.

IPCC. *Climate Change 2007: The Physical Science Basis. Contribution of Working Group I to the Fourth Assessment Report of the Intergovernmental Panel on Climate Change.* Cambridge: Cambridge University Press, 2007a.

IPCC. *Climate Change 2007: Impacts, Adaptation and Vulnerability. Contribution of Working Group II to the Fourth Assessment Report of the Intergovernmental Panel on Climate Change.* Cambridge: Cambridge University Press, 2007b.

IPCC. *Climate Change 2007: Mitigation of Climate Change. Contribution of Working Group III to the Fourth Assessment Report of the Intergovernmental Panel on Climate Change.* Cambridge: Cambridge University Press, 2007c.

IPCC. *Climate Change 2007: Synthesis Report. Contribution of Working Groups I, II and III to the Fourth Assessment Report of the Intergovernmental Panel on Climate Change.* Pachauri, R. K., and Reisinger, A. (Eds.). Geneva, Switzerland, 2007d.

Irwin, William. *The New Niagara.* University Park: Pennsylvania State University Press, 1996.

Jackson, Donald C. *Building the Ultimate Dam.* Lawrence: University of Kansas Press, 1995.

Jackson, Kenneth T. *Crabgrass Frontier.* New York: Oxford University Press, 1985.

Kay, Jane Holtz. *Asphalt Nation.* Berkeley: University of California Press, 1997.

Kirsch, David. *The Electric Vehicle and the Burden of History.* New Brunswick, N.J.: Rutgers University Press, 2000.

Kolbert, Elizabeth. *Field Notes from a Catastrophe: Man, Nature, and Climate Change.* New York: Bloomsbury, 2006.

Krauss, Clifford, and Jad Mouawad. "Oil Industry Backs Protests of Emissions Bill." *New York Times,* August 18, 2009. http://www.nytimes.com/2009/08/19/busi ness/energy-environment/19climate.html.

Kuhn, Thomas S. *The Structure of Scientific Revolutions,* 3rd ed. Chicago: University of Chicago Press, 1996.

Lamb, H. H. *Climate, History and the Modern World,* 2nd ed. London: Routledge, 1995.

Lamb, Simon, and David Sington. *Earth Story: The Shaping of Our World.* Princeton, N.J.: Princeton University Press, 1998.

Landes, David. *The Unbound Prometheus: Technological Change and Industrial Development in Europe.* New York: Cambridge University Press, 1969.

Leopold, Aldo. *A Sand County Almanac, and Sketches Here and There.* [1948.] New York: Oxford University Press, 1987.

Lewis, Tom. *Divide Highways.* New York: Penguin Books, 1997.

Linden, Eugene. *The Winds of Change: Climate, Weather, and the Destruction of Civilizations.* New York: Simon & Schuster, 2006.

Loeb, P. *Moving Mountains: How One Woman and Her Community Won Justice from Big Coal.* Lexington: The University Press of Kentucky, 2007.

Lomborg, Bjørn. *The Skeptical Environmentalist: Measuring the Real State of the World.* Cambridge: Cambridge University Press, 2001.

Lomborg, Bjørn. *Cool It: The Skeptical Environmentalist's Guide to Global Warming.* New York: Vintage, 2008.

Lovins, Amory. *Soft Energy Paths.* New York: HarperCollins, 1979.

Maher, Neil. "Neil Maher on Shooting the Moon." *Environmental History* 9 no. 3 (2004): 12 pars. 27 Jan. 2006. http://www.historycooperative.org/cgi-bin/justtop. cgi?act=justtop&url=http://www.historycooperative.org/journals/eh/9.3/maher.html.

Mann, Michael E., and Lee R. Kump. *Dire Predictions: Understanding Global Warming, The Illustrated Guide to the Findings of the IPCC.* London: DK Publishing, 2009.

Marsh, G. P. *Man and Nature.* Cambridge: The Harvard University Press, 1965. (This is an annotated reprint of the original 1864 edition.)

Martin, Albro. *Railroads Triumphant: The Growth, Rejection and Rebirth of a Vital American Force.* New York: Oxford University Press, 1992.

Maslin, Mark. *Global Warming: A Very Short Introduction.* Oxford: Oxford University Press, 2004.

McGreevy, Patrick V. *Imagining Niagara.* Amherst: University of Massachusetts Press, 1994.

McHarg, Ian. *Design with Nature.* New York: John Wiley and Sons, 1992.

McKinsey, Elizabeth. *Niagara Falls: Icon of the American Sublime.* Cambridge: Cambridge University Press, 1985.

McNeil, John R. *Something New Under the Sun: An Environmental History of the Twentieth-Century World.* New York: Norton, 2001.

McShane, Clay. *Down the Asphalt Path,* New York: Columbia University Press, 1994.

Melosi, Martin. *Coping with Abundance.* New York: Knopf, 1985.

Melosi, Martin. *Sanitary City.* Baltimore: Johns Hopkins University Press, 1999.

Merchant, Carolyn. *Major Problems in American Environmental History.* New York: Heath, 2003.

Merrill, Karen. *The Oil Crisis of 1973–1974.* New York: Bedford, 2007.

Miller, B. *Coal Energy Systems.* Burlington, Mass.: Elsevier Academic Press, 2005.

Mitchell, J. G. "When Mountains Move." *National Geographic,* March 2006. http://www7.nationalgeographic.com/ngm/0603/feature5/index.html.

Mokyr, Joel. *Twenty-five Centuries of Technological Change.* New York: Harwood Academic Publishers, 1990.

Mokyr, Joel, ed. *The Economics of the Industrial Revolution.* Totowa, N.J.: Rowman & Allanheld, 1985.

Montrie, Chad. *To Save the Land and People: A History of Opposition to Surface Coal Mining in Appalachia.* Chapel Hill: University of North Carolina Press, 2003.

Mooney, Chris. *Storm World: Hurricanes, Politics, and the Battle Over Global Warming.* New York: Mariner Books, 2008.

Moorhouse, John C., ed. *Electric Power: Deregulation and the Public Interest.* San Francisco: Pacific Research Institute for Public Policy, 1986.

Motavalli, Jim. *Forward Drive: The Race to Build "Clean" Cars for the Future.* San Francisco: Sierra Club Books, 2001.

Mumford, Lewis. *Technics and Civilization.* New York: Harcourt, 1963.

Nash, Roderick. *Wilderness and the American Mind.* New Haven, Conn.: Yale University Press, 1982.

Novak, Barbara. *Nature and Culture.* New York: Oxford University Press, 1980.

Nye, David. *Technological Sublime.* Boston: MIT Press, 1996.

Nye, David. *Electrifying America.* Boston: MIT Press, 1999.

Ocean Studies Board. *Abrupt Climate Change: Inevitable Surprises.* Washington D.C.: National Academy Press, 2002.

Oliens, Roger M., and Dianna Davids. *Oil and Ideology: The American Oil Industry, 1859–1945.* Chapel Hill: University of North Carolina Press, 1999.

Opie, John. *Nature's Nation.* New York: Harcourt Brace, 1998.

Parker, Ann. "The Siren Call of the Seas: Sequestering Carbon Dioxide." *Science and Technology Review* (May, 2004) (Lawrence Livermore National Laboratory). https://www.llnl.gov/str/May04/Caldeira.html.

Pearce, Fred. "'Climategate' was PR Disaster that Could Bring Healthy Reform of Peer Review." *The Guardian,* February 9, 2010. http://www.guardian.co.uk/environment/2010/feb/09/climate-emails-pr-disaster-peer-review.

Perkins, J. H. *Geopolitics and the Green Revolution: Wheat, Genes, and the Cold War.* New York: Oxford University Press, 1997.

Pinchot, Gifford. *Breaking New Ground.* New York: Island Press, 1998.

Pollan, Michael. *Second Nature.* New York: Delta, 1992.

Pollan, Michael. *Omnivore's Dilemma.* New York: Penguin, 2007.

Ponte, Lowell. *The Cooling.* London: Prentice Hall, 1976.

Pyne, Stephen. *Fire in America.* Princeton, N.J.: Princeton University Press, 1982.

Rabe, Barry G. *Statehouse and Greenhouse: The Emerging Politics of American Climate Change Policy.* Washington, D.C.: Brookings Institution Press, 2004.

Reese, E. *Lost Mountain: A Year in the Vanishing Wilderness: Radical Strip Mining and the Devastation of Appalachia.* New York: Penguin Group, 2006.

Reuss, Martin. *Water Resources Administration in the United States: Policy, Practice, and Emerging Issues.* Ann Arbor: Michigan State University Press, 1993.

Rifkin, Jeremy. *The Hydrogen Economy.* New York: Penguin, 2003.

Roberts, Peter. *Anthracite Coal Communities.* 1904. Westport, Conn.: Greenwood Publishers, 1970.

Rottenberg, Dan. *In the Kingdom of Coal: An American Family and the Rock That Changed the World.* New York: Routledge, 2003.

Roy, Andrew. *The Coal Mines.* New York: Robison, Savage & Co., 1876.

Ruse, Michael. *The Evolution Wars: A Guide to the Debates.* New Brunswick, N.J.: Rutgers University Press, 2002.

Sabin, Paul. *Crude Politics: The California Oil Market, 1900–1940.* Berkeley: University of California Press, 2005.

Samuelsohn, Darren. "Top U.N. Climate Diplomat Announces Resignation." *New York Times,* February 18, 2010. http://www.nytimes.com/cwire/2010/02/18/18climatewire-top-un-climate-diplomat-announces-resignatio-89589.html.

Schiffer, Michael B., Tamara C. Butts, and Kimberly K. Grimm. *Taking Charge: The Electric Automobile in America.* Washington, D.C.: Smithsonian Institution Press, 1994.

Segal, Howard P. *Technological Utopianism in American Culture,* 20th Anniv. Ed. Syracuse, NY: Syracuse University Press, 2005.

Sheppard, Muriel. *Cloud by Day: The Story of Coal and Coke and People.* Pittsburgh: University of Pittsburgh Press, 2001.

Solomon, Lawrence. *The Deniers: The World Renowned Scientists Who Stood Up Against Global Warming Hysteria, Political Persecution, and Fraud.* Minneapolis: Richard Vigilante Books, 2008.

Smil, Vaclav. *Energy in China's Modernization: Advances and Limitations.* Armonk, N.Y.: M. E. Sharpe, 1988.

Smil, Vaclav. *Energy in World History.* Boulder, Colo.: Westview Press, 1994.

Smith, Duane. *Mining America: The Industry and the Environment, 1800–1980.* Lawrence: Kansas University Press, 1987.

Stearns, Peter N. *The Industrial Revolution in World History.* Boulder, Colo.: Westview Press, 1998.

Steinberg, Theodore. *Nature Incorporated: Industrialization and the Water of New England.* New York: Cambridge University Press, 1991.

Stevens, Joseph E. *Hoover Dam.* Norman: University of Oklahoma Press, 1988.

Stilgoe, John R. *Metropolitan Corridor: Railroads and the American Scene.* New Haven, Conn.: Yale University Press, 1983.

Stradling, David. *Smokestacks and Progressives: Environmentalists, Engineers, and Air Quality in America, 1881–1951.* Baltimore: Johns Hopkins University Press, 1999.

Tarbell, Ida. *All in the Day's Work: An Autobiography.* Champaign: University of Illinois Press, 2003.

Tarr, Joel. *The Search for the Ultimate Sink.* Akron, Oh.: University of Akron Press, 1996.

Tarr, Joel, ed. *Devastation and Renewal.* Pittsburgh: University of Pittsburgh Press, 2003.

Trachtenberg, Alan. *Incorporation of America.* New York: Hill and Wang, 1982.

Van Andel, and H. Tjeerd. *New Views on an Old Planet: A History of Global Change,* 2nd ed. Cambridge: Cambridge University Press, 1994.

Velasquez-Manoff, Moises. "The Tiny, Slimy Savior of Global Coral Reefs?" *The Christian Science Monitor,* February 6, 2009. http://features.csmonitor.com/environment/2009/02/06/the-tiny-slimy-savior-of-global-coral-reefs/.

Victor, David G. *The Collapse of the Kyoto Protocol and the Struggle to Slow Global Warming.* Princeton, N.J.: Princeton University Press, 2001.

Vlasic, Bill. "As Market Shifts, Ford Sees Profit Fleeing." *New York Times,* August, May 23, 2008. http://www.nytimes.com/2008/05/23/business/23ford.html.

Weart, Spencer. *The Discovery of Global Warming,* Rev. Exp. Ed. Cambridge: Harvard University, 2008.

Williams, Michael. *Americans and Their Forests.* New York: Cambridge University Press, 1992.

Yergin, Daniel. *The Prize: The Epic Quest for Oil, Money & Power.* New York: Free Press, 1993.

## WEB SOURCES

EPA site on Global Warming. http://www.epa.gov/climatechange/.

CA Climate. http://www.climatechange.ca.gov/policies/index.html.

Cizik. On the Care of Creation. http://www.creationcare.org/resources/declaration.php.

Climate Change and Cities Initiative (CCCI). http://www.unhabitat.org/categories.asp?catid=550.

Cornell University. http://www.cals.cornell.edu/cals/public/comm/pubs/ecalsconnect/vol14–2/features/carbon-footprint.cfm.

"The CRU Hack." *RealClimate,* November 20, 2010. http://www.realclimate.org/index.php/archives/2009/11/the-cru-hack/.

DOE (Department of Energy). http://www.energy.gov/about/index.htm.

Envirolink. http://www.envirolink.org/topics.html?topicsku=2002109190933& topic=Climate%20Change&topictype=topichttp://www.cln.org/themes/ global_warming.html.

Faulk, Richard, and John S. Gray. "Getting the Lead Out: The Misuse of Public Nuisance Litigation by Public Authorities and Private Counsel." (2006). http:// works.bepress.com/richard_faulk/21//.

Ford Motor Co. http://biz.yahoo.com/nytimes/080522/1194777767445.html?.v=4.

"Ford Reveals Small-Car Vision for North America." http://media.ford.com/article_ print.cfm?article_id=27479.

General power generation. http://www.ala.org/ala/acrl/acrlpubs/crlnews/backissues 2005/january05/alternateenergy.cfm.

Gingrich, Newt. "The Gingrich-Pelosi Climate Change Ad: Why I Took Part." http:// newt.org/tabid/193/articleType/ArticleView/articleId/3351/Default.aspx.

Gore, Alternatives: http://thinkprogress.org/gore-nyu/.

"Gore, Gingrich Face off on Global Warming Bill." MSNBC.com. http://www. msnbc.msn.com/id/30386828/.

Hansen, James. "James Hansen's Open Letter to Obama: From a Top NASA Scientist and a Fellow Parent." Huffington Post. http://www.huffingtonpost.com/2009/ 01/05/james-hansens-open-letter_n_155199.html.

Henderson, Caspar. "Reason and Light." *New Statesmen,* May 15, 2006. http://www. newstatesman.com/200605150065.

Inhofe, James M. Floor Speeches. http://inhofe.senate.gov/pressreleases/climate. htm.

Intergovernmental Panel on Climate Change. http://www.ipcc.ch/.

International Commission on Stratigraphy. http://www.stratigraphy.org/.

International Federation of Red Cross. "Climate Change Adaptation Strategies for Local Impact: Key Messages for UNFCCC Negotiators." http://www.unfccc.int/ resource/docs/2009/smsn/igo/054.pdf.

KLM Auto. http://klmperformance.com/automotive-news/.

Moyers, Bill. http://www.pbs.org/moyers/journal/08152008/transcript1.html.

Nobel Prize. http://nobelprize.org/nobel_prizes/peace/laureates/2007/press.html.

North American Leaders' Declaration. http://www.whitehouse.gov/the_press_office/ North-American-Leaders-Declaration-on-H1N1/.

The Oregon Petition. Global Warming Petition Project: http://www.petitionproject. org/index.php/.

Pew Center on Global Climate Change. http://www.pewclimate.org/what_s_being_ done/in_the_states/regional_initiatives.cfm.

Schumacher, E. F.: Technology with a Human Face. http://www.cooperativeindivid ualism.org/schumacher_technology_with_human_face.html.

Science and Environmental Policy Project: http://www.sepp.org/policy%20decla rations/home.html.

Secondary Energy Infobook. www.need.org/needpdf/Secondary%20Energy%20 Infobook.pdf.

Simon, Stephanie. "Cities, Towns Step Up Global Warming Fight." January 3, 2007. http://www.stopglobalwarming.org/sgw_read.asp?id=41348132007

Snowball Earth. http://www.snowballearth.org/.

Thatcher, Margaret. Speech at the 2nd World Climate Conference, November 6, 1990. Margaret Thatcher Foundation. http://www.margaretthatcher.org/.

UN_HABITAT. http://www.unhabitat.org/content.asp?typeid=19&catid=570&cid=6003.

United Nations Programs on Climate Change. http://unfccc.int/2860.php for younger readers: http://www.fao.org/kids/en/gw-un.html; http://cyberschoolbus.un.org/treaties/global.asp.

UN-REDD Programme. http://www.un-redd.org/Home/tabid/565/language/en-US/Default.aspx.

Weart, Spencer. "A Personal Note." http://www.aip.org/history/climate/SWnote.htm.

Weart, Spencer. *The Discovery of Global Warming*. http://www.aip.org/history/climate/. Weart interview. http://voices.washingtonpost.com/capitalweathergang/2009/11/perspective_on_a_climate_scien.html.

# Index

## About the Authors

BRIAN C. BLACK is professor of history and environmental studies at Penn State Altoona, Altoona, PA. He is the author of several books and essays including *Petrolia: The Landscape of America's First Oil Boom*. He specializes in the environmental history of North America, particularly energy.

GARY J. WEISEL is associate professor of physics at Penn State Altoona, Altoona, PA. He writes on the history of 20th-century physical science communities and also does research in nuclear physics and materials science.